OCEAN LINERS

OCEAN LINERS

A New History

Anthony Burton

PEN & SWORD
TRANSPORT
AN IMPRINT OF PEN & SWORD BOOKS LTD.
YORKSHIRE – PHILADELPHIA

First published in Great Britain in 2024 by
Pen and Sword Transport
An imprint of
Pen & Sword Books Ltd.
Yorkshire - Philadelphia

ISBN 978 1 39904 979 5

A CIP catalogue record for this book is available from the British Library.

Typeset in 10/13 pt Palatino by SJmagic DESIGN SERVICES, India.

Printed and bound in the UK on paper from a sustainable source by CPI Group (UK) Ltd., Croydon. CR0 4YY.

Pen & Sword Books Ltd. incorporates the imprints of Pen & Sword Books: After the Battle, Archaeology, Atlas, Aviation, Battleground, Discovery, Family History, History, Maritime, Military, Politics, Select, Transport, True Crime, Fiction, Frontline Books, Leo Cooper, Praetorian Press, Seaforth Publishing, Wharncliffe and White Owl.

For a complete list of Pen & Sword titles please contact

PEN & SWORD BOOKS LIMITED
George House, Units 12 & 13, Beevor Street, Off Pontefract Road, Barnsley, South Yorkshire, S71 1HN, England
E-mail: enquiries@pen-and-sword.co.uk
Website: www.pen-and-sword.co.uk

or

PEN AND SWORD BOOKS
1950 Lawrence Rd, Havertown, PA 19083, USA
E-mail: uspen-and-sword@casematepublishers.com
Website: www.penandswordbooks.com

FSC www.fsc.org MIX Paper | Supporting responsible forestry FSC® C013604

Contents

Introduction

What distinguishes the ocean liner from any other ship? The criteria it must meet to qualify for this book are quite simple. First, and most obviously, it must cross an ocean, not just hop across the English Channel, for example. Secondly, although it might carry small amounts of cargo, the main concentration must be on passengers. Thirdly, it must offer a regular service between its principal ports, without offering special arrangements for excursions, to distinguish it from cruising ships. Provided all these requirements are met, then the vessel qualifies. It makes no difference whether it is moved by sail, steam or diesel engine.

The golden age of the ocean liner: the Cunard liner *Berengaria* at New York.

Ships have fascinated me since I was a boy, partly because there was family involvement in the industry. My great grandfather was one of three Riley brothers, who built marine boilers at their works in Stockton-on-Tees. It was only years later that I tracked any down. One was installed in the paddle steamer *Kingswear Castle*, but was recently replaced, but the other in the steam herring drifter *Lydia Eve* is still there and still regularly put to use in the preserved vessel. None, alas, would ever find its way into anything as grand as an ocean liner.

The other question is, why ocean liners rather than any other type of ship? The answer is simple; at the best, they are among the most beautiful vessels ever built. They also offer a curious dichotomy. The fundamental design would be based on the very latest technology and aimed at maximum efficiency. The interiors would represent current fashions in interior design. They also act as microcosms of the societies they served, ranging from high luxury for those fortunate enough to travel first class, and often the barest minimum of comfort for those, often immigrants, travelling steerage. In other words, the ocean liner is unique.

Chapter One

A Scheduled Service

Until the early years of the nineteenth century, passengers wishing to make ocean voyages, such as that between Britain and America, simply had to take whatever ship was available with such accommodation as it might be able to provide, which all too often was basic, cramped and uncomfortable. Having got on board there was no guarantee that the ship would head straight off for its destination but might well call in at other ports to pick up extra cargo. Even a direct voyage was a lengthy business. Marc Brunel, who had escaped Revolutionary France to make a home in America, left New York for England in 1799 and arrived at Falmouth 28 days later, which was considered a reasonably fast passage. No one expressed any great surprise if a voyage to the west took three months. All that was about to change.

In 1817, an American consortium set up the Wright, Thompson, Marshall & Thompson Line in association with Cropper, Benson & Co. of Liverpool. The company flag featured a black circle, which led to the name being eventually reduced to the more manageable Black Ball Line. In October of that year, an announcement appeared in the New York *Evening Post* informing the public that four ships were available, the *Pacific, Amity, Courier* and *James Monroe*. 'It is the intention of the owners that one of these vessels shall sail from New York on the 5th of each month and one from Liverpool on the 1st of each month.' Services got under way the following year. These were fine fully-rigged ships, with sails carried on three masts and passages were notably faster than those by other vessels of the time. The average times between New York and Liverpool were 23 days travelling east and 40 days westward. The best times recorded were just over 15 days, with one remarkable voyage, when the Black Ball *Columbia* managed a westward journey in less than half the usual time, reaching Liverpool in 15 days 23 hours. By 1822, the company was offering two sailings a month at a cost of 35 guineas, over £3,000 at today's prices. That is a lot of money and those who paid it would expect good value, but did they get it?

The first vessel to leave New York was the *James Monroe*, a 424 ton vessel with two decks. Passengers got what was called a stateroom, rather a grand name for a space just 8ft square, but panelled with satinwood and entered through an archway flanked by marble columns. The box beds had hair mattresses and though hardly en

The Black Ball packet *New York* off Alisa Craig in the Firth of Forth by William Clark. The distinctive 'black ball' motif can be seen on the fore lower topsail.

suite, there was a small basin for washing and a chamber pot. Meals were taken in the saloon, a much grander affair, highly decorated with horsehair chairs and a mahogany table. The real luxury, however, was the availability of fresh food. Bread was baked daily, and up on deck was livestock, chickens for eggs, cows for milk and pigs and sheep that were not going to be alive by the end of the voyage. By the standards of later transatlantic liners, the facilities were primitive, but compared to those offered by other ships of the day they were luxurious.

William Cobbett, the English radical, best known for his book *Rural Rides*, spent some time in America and wrote a guide book for emigrants in 1829, in which he was fulsome in his praise of American packets.

> The ship will be no other than an American one, if you wish for a *quick* and *safe* passage. The Americans sail *faster* than others, owing to the greater *skill* and greater *vigilance* of their captains, and to their greater sobriety and the wise rule that they observe with regard to other men. They carry *more sail* than other ships, because the captain is everlastingly looking out. I have crossed the Atlantic three times, once in an English merchant ship, once in a king's ship, and once in a king's packet; and I declare, that the superiority of the Americans is

> indeed so decisive, that, if I were going to cross again, nothing should prevail on me to go abroad of any ship but an American one. I never knew an American captain to *take off his clothes to go to bed, during the whole voyage*; and I never knew any other who did not do it. The consequence of this great watchfulness is that advantage is taken of every puff of wind, while the risk from squally and sudden gales is, in a great measure, obviated.

The eulogy is a little surprising as he had endured an unfortunate experience with the Black Ball line. He had planned to leave New York on the *Amity* but word got around that in his luggage were the remains of Thomas Paine, author of *Rights of Man,* considered the most outrageously revolutionary book of the age. Cobbett wanted the bones interred and honoured back in Britain, but the passengers who had booked for the same ship, objected to having either Cobbett or the dead Paine as their companions. The line refused to sell him a ticket, but he did manage to get a berth on the *Hercules* that left America on 30 October 1819, arriving with Paine's bones in Liverpool on 21 November.

The success of the Black Ball line encouraged others to set up similar systems, but there was a limit to the number of comparatively rich men and women who wanted to cross the Atlantic. The Black Ball turned to a new trade – immigration.

The great advantage of this trade was that there was no need to provide any luxuries whatsoever and the ship owners could cram in as many bodies as they could with no regard for comfort. The trade grew steadily. Statistics from a British government report on immigration, produced in 1873, show that 45,000 left Britain in 1835, but by 1852 the number had risen to 369,000, of whom the majority came from Ireland, though a large number also crossed the Atlantic from Germany. The main destination was the United States, followed by Canada. There was also a large number heading for Australia and New Zealand, though never as many as left for the Americas. Rather confusingly, most Australian migrants were carried by a Liverpool based line, also called Black Ball. The conditions under which the immigrants travelled ranged from barely tolerable to appalling. The passengers who paid for a berth might have imagined that they were getting some sort of bunk to themselves but often found that instead they had a space some six-foot square into which they had to squeeze with three others. On some ships they were required to bring their own food for the voyage, while on others the ship provided whatever food they thought was the bare minimum to keep their passengers alive. Water rations were handed out, but from barrels that were often tainted with whatever had been previously stored in them. Washing and toilet facilities were primitive. Stephen de Vere, a philanthropist who chose to travel steerage to see the conditions for himself, gave evidence to the House of Lords Committee on Emigration of 1847:

> Hundreds of poor people, men, women, and children of all ages, from the drivelling idiot of ninety to the babe just born, huddled together without light, without air, wallowing in filth and breathing a fetid atmosphere, sick in body, dispirited in heart, the fevered patients lying between the sound, in sleeping places so narrow as almost to deny them the power of indulging, by a change of position, the natural restlessness of the disease … living without food or medicine, except as administered by the hand of casual charity, dying without the voice of spiritual consolation, and buried in the deep without the rites of the church.

Many of the ships were of poor quality, but even life on the finest vessels could be made miserable by the actions of the captain and his officers. Vere Foster was a British educationalist and humanist, who devoted much of his life working to improve the living conditions and educational opportunities for the Irish poor. He travelled on the *Washington*, a notably well-built vessel, that carried some 900 passengers. Once at sea, it was clear that the poor passengers were being treated with total contempt. Each was supposed to get a daily ration of water, but after just a few had been served, the mate told the rest that no more would be issued. Foster remonstrated with the mate and was told to mind his own business or he would be knocked down. He had just started to write a letter of complaint to the captain when he met the mate again, and this time he was knocked flat. Eventually, the letter was written and presented, and all the answer he got was to be abused, called a 'damned pirate' and told that if he made any more trouble he would be kept in irons for the rest of the journey. The abuses continued, with food not being provided for days on end, and passengers often viciously assaulted on the least excuse. At the end of the voyage, Foster tried to get action taken against the captain and his officers but met with official indifference; it was Foster's word against the captain's and nothing could be done.

The life of the ordinary sailors on board these vessels was hardly better than that of the passengers in steerage. They were subjected to regular beatings for any infringement of the rules – or simply for not being quite quick enough to respond to an order. Yet even the most critical captain who was happy to announce that regular beatings were the only way to keep discipline on board had to admit that it was their willingness to keep the ship seaworthy in even the worst conditions that demanded at least grudging respect. Just how dangerous the work could be can be illustrated by the experience of that famous travel writer Eric Newby in his book *The Last Grain Race.* Sailing ships were still at work in 1938 when he joined a four-masted barque as an apprentice for a voyage from Europe to Australia. They were hit by a fierce storm, and all hands were ordered aloft to take in sail. Newby lost his grip but was caught by the wind and blown backwards from the yardarm on which he had been working all the way to the rigging of the mast behind him, where he was able to cling on and save himself. He was immediately ordered back up to carry on working. Many a sailor on a square-rigged ship was not that lucky.

Many of the immigrant ships never survived the passage at all. Between 1847 and 1853, fifty-nine of these vessels heading to America were lost. The death toll among the passengers was terrible. When the *Exmouth* was wrecked on a voyage from Londonderry to Quebec for example, only three of the 251 passengers aboard survived. One particularly tragic incident occurred when the *Powhattan* was wrecked off the coast of New Jersey in 1851, within sight of land. Those on board could hear shouted suggestions from people on shore, but no one made it out of the vessel alive as she slowly broke up. After 1850, the migrant numbers began to drop and conditions improved.

Conditions for paying passengers in the packets were a world away from those of the unfortunates in the immigrant ships or tucked away in steerage. They had state rooms, usually with two bunks. A solitary passenger who valued privacy could, however, pay extra and have the cabin to themselves. The ships offered a great deal in the way of comforts, including a saloon where they could relax, and on some vessels a separate smoking room for the men. One vessel, the *Liverpool,* even offered the luxury of a bathroom, though all it had to offer was a salt water shower. Meals on board were gargantuan feasts, served four times a day. Tyrone Power, an Irish comedian not the Hollywood actor, described the breakfast menu of 'boiled ham, spatch-cock, eggs, frizzled bacon and mutton cutlets'. At the end of the meal, he had a choice between gin or brandy. It must also have pleased the Irishman that Guinness was also available on board.

Between meals there were games of cards or dominoes in the saloon or shuffle board on deck. Presiding over it all was the captain and to be invited to dine at the captain's table was an honour. The attitude of the ship's officers to the passengers was exactly the opposite of that shown to the miserable immigrants. They were treated as guests and would never be referred to as 'pirates' or threatened with being put in irons. In return, the captain was usually generously rewarded by the passengers. On one notable occasion, the *Yorkshire,* the fastest of all the sailing packets, had among its passengers a troupe of Viennese dancers, who on reaching New York in record time, performed the dance to Delibes' *Pas des Fleurs* on the quarterdeck for him and then presented him with a silver pitcher.

If all this makes a voyage sound like a riot of overindulgence and pleasure, then one has also to remember that this was an Atlantic crossing and was subject to two miseries – storms and calms. The former brought for many the horrors of sea sickness. Sarah Mytton Maury was at first delighted with the motion of the ship:

> Truly it was unlike all other motions that I had ever experienced in a ship; prolonged, and breathing, and swelling, while the vessel, plunging gently onwards, seemed to recognise with joy the friendly welcome of the ocean billow … for the first time I felt and understood the *rapture of the sea*.

Then a gale rose and rapture turned to misery. 'Soon I became sick and sore, a dead-alive wretch, neither hearing nor seeing, nor knowing anything around me. The good steward managed to drag me to my state room, and here I lay for thirty-seven hours in mortal sickness.' For sufferers from sea sickness, becoming becalmed no doubt came as a welcome relief. For everyone else it was simply a period of intense frustration as they were like Coleridge's ancient mariner: 'as idle as a painted ship upon a painted ocean'. Tempers frayed. Harriet Martineau at least found it amusing to observe her fellow passengers:

> After three days of calm, there may be heard a subdued tone of scolding from the whist party at the top of the table, and a stray oath from some check-mated person lower down; and while the ladies are brushing their hair in the cabin, certain items of information are apt to be given of how Mrs. A looked when the lady's partner turned up trumps, and how shockingly Mr. B pushed past Mr. C in going up to the cabin to dinner.

There was nothing to be done about calms and storms, but it was in everyone's interest to get through a voyage as quickly as possible; the passengers were spared potential extra days of misery and the owners made more money. One answer was to improve the design and construction of the ships. Rees's *Cyclopaedia* published between 1819 and 1820 had a section on ship building that made it plain that when it came to the science of shipbuilding, very little had changed for centuries.

> The scientific part of ship-building has been too much neglected; and although some few years have elapsed since mathematicians (particularly in France) have laboured with some success, yet their discoveries are so much enveloped in profound calculations, that shipbuilders, in general, have scarcely been able to derive any advantage from them.

The main argument was between building ships of traditionally broad beam or changing to a narrow. Rees offered no real answer:

> A long, narrow ship will require less sail to gain velocity, consequently the masts will be lower, the rigging lighter, and the vessel navigated by fewer hands. On the contrary, a ship's being broader at the line of flotation will admit of being narrower on the floor, particularly at the fore and after parts; that by being broader it can carry more sail, and more readily rise upon the waves than a narrow one.

In America, however, narrow beamed vessels with sharply raked bows had been developed during the war with Britain of 1812 that were either schooner rigged, with fore and aft sails, or brigs square rigged with two masts. They proved to be faster than the British warships and when the war ended, the design may not have been suitable for cargo ships, but was ideal for fast packets. They became known as Baltimore clippers, although they did not have the distinctive hollow bows of true clippers. In 1832, the first of a new generation of clippers appeared with the *Anne McKim,* square rigged and three masted. The American clippers were the fastest on the Atlantic crossings and a few British yards started following their lead, notably the Scott yard at Greenock on the Clyde. They took a genuinely new approach, testing hull designs using models, at first in the still waters of a dock, but later on Loch Thom, where the conditions were in scale with those met by the real ships at sea.

One area where there had been scientific progress in the eighteenth century was in navigation. It might seem obvious that if you have a compass and a chart, all you have to do is point the vessel in the right direction, and if you know how fast you are going, then you can tell where you are at any time, but life at sea is not that simple. Even if a vessel didn't have to tack to take advantage of the wind, it would not necessarily be going in the way the compass indicated as it could be drifting to one side or the other, moved by wind and tide. Speed had been determined for a long time by the simple device of the log – which might have originated in a piece of actual wood. Basically, it

The American clipper *Great Republic*, from the *Illustrated London News*, 1855.

is a floating device attached to a rope with knots at specific intervals. The log is thrown overboard, and remains afloat on the spot where it landed, and as the ship moves away, the rope is paid out and the number of knots counted for a time determined by an hour glass. This gives the speed in nautical miles per hour, or knots, which can then be entered in the log book. Latitude can be determined by measuring the height of the sun above the horizon at noon, a measurement made more accurate by the introduction of the sextant in 1757.

Finding longitude presented a different problem. For navigational purposes the earth is divided into 360 degrees of longitude, and as it rotates it moves through one degree in four minutes. If the navigator knows when noon was at a particular degree of longitude, then the difference between that and noon where he is will give him the longitude. The difficulty was finding a clock that could keep accurate time despite being rolled around at sea. The problem was solved by John Harrison, who produced his first accurate chronometer in 1735 and continued to develop it over the next two decades. The standard longitude was set at 0° at the point where the imaginary line would pass through the Greenwich observatory. By 1767, the first Nautical Almanac had appeared, making it possible not only to find positions in relation to the sun, but also by observing, moon, stars and planets and measuring their height at a given time. I was still being taught to

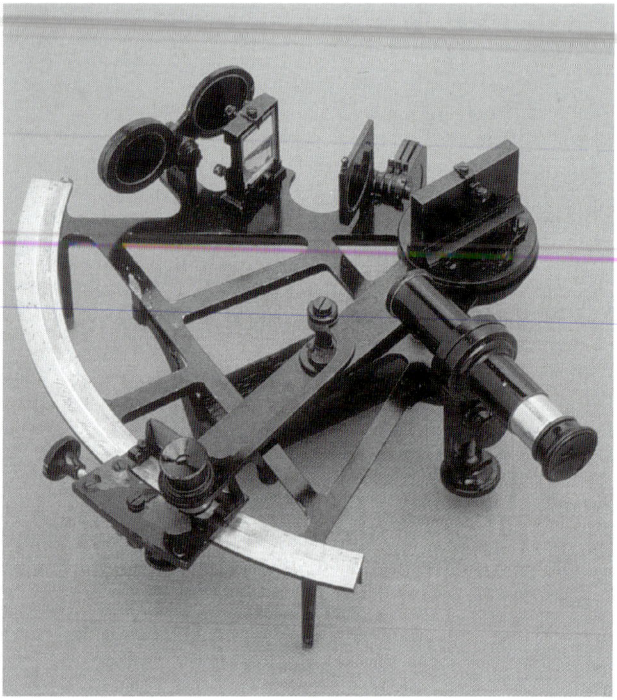

A sextant, first used to measure the angle between the sun and the horizon at noon to determine the ship's latitude. Later it could also be used to make astronomical observations as well.

fix a position using a sextant and astronomical observations when qualifying as a navigator in the 1950s. All this made sailing safer, but measurements depended still on clear skies – and navigational errors remained a hazard.

The new fast clippers and improved navigational aids made Atlantic crossings safer than they had been in earlier times, but sinkings were still all too common. Ships might be improved, but no one could change the weather – storms could wreak havoc with sailing ships at sea. There was always the possibility of collisions, especially in fog, when another ship might suddenly appear through the gloom, too late for the watch to give warning.

It was not just the Atlantic, however, that was regularly crossed by passenger ships in the age of sail. Once the British began to colonise India, there was a regular call for a passenger service to take out the officers and families of the East India Company. The company's ships, known as East Indiamen, carried cargo and passengers and were armed with cannon in the late eighteenth century. By the end of the century, they had largely abandoned carrying armaments, not because the voyages were any safer, but simply because they needed the space in order to compete with rivals This became even more necessary when the Company lost its monopoly on the Indian trade. *Neptune*, shown here, is a typical East Indiaman of the later period, when heavy armament was no longer carried.

The major change in ship design came in 1789, when the company decided they needed bigger ships, and soon they had an impressive fleet of thirty-six 1,200 tonners and forty 800 tonners, and the vessels were expected to make a grand total of six voyages. Later, the ships were often built in India, where there were ample supplies of first-class timber. They might have been more successful if it were not for the curious system used for appointing commanders. Being in command of an East Indiaman was a highly profitable business and the man in charge could make a fortune, as much as £10,000 a voyage. Needless to say, competition for the post was fierce, but instead of appointing the

The East Indiaman *Neptune* off the south coast of England by Thomas Luny. She is shown in two aspects: sailing towards and away from the viewer.

most competent man, it went to tender, and the one who offered the Company the most money got the post. Considerable wealth does not necessarily go with superior seamanship.

The crew of East Indiamen at the end of the eighteenth century often consisted of lascars, a name given to seamen from India and Malaya. Partly, this was a simple economic decision – they worked for very little pay – and also due to a shortage of British sailors. The lascars were often ill-prepared for the work, lived on board in wretched conditions, and if they proved too ill to work, they were set ashore as soon as the vessel arrived in Britain and simply abandoned to their fate. The Company, however, did insist on looking after their spiritual welfare when on board; Commanders were under strict instructions to hold services every Sunday, though as the crew were probably either Hindu or Muslim, it could hardly have brought them much comfort.

In the early days, passengers generally boarded the vessel when she was anchored off the south coast of England. They would be rowed out and usually charged 5 shillings for the trip, but it could cost as much as £5 in really bad weather. But that was very little compared with the cost of the voyage. Most of the passengers were going out as officers of the Company, and they were charged according to their rank, from £95 for a subaltern to £235 for a general, but anyone wanting the best cabin on board could expect to shell out perhaps 400 guineas. They were expected to provide their own wine and servants but were allowed to bring a massive amount of luggage aboard, up to 3½tons for the wealthiest. According to R.C. Leslie, in his book *Old Sea Wings* (1894), the passengers had a menagerie on board to keep them fed:

> The heavy long-boat of even a fast 'line-of-packet' ship … looked more like a working model of Noah's Ark than anything likely to save life at sea, or even to live upon it … the boat, as she lay upon the ship's deck, was full of live provender; being divided as to her lower hold into pens for sheep and pigs; while upon the first floor, or main-deck, quacked ducks and geese, and above them (literally in the cock-loft) were coops for another kind of poultry.

The East Indiamen, however, were as prone to disaster as every other large sailing ship of the time. The *Euphrates*, for example, is a fine-looking vessel. Built at Liverpool in 1834, she made her first voyage to Calcutta in 1836-7. In 1868, however, she was on her way back to Liverpool with a cargo of palm nuts and oil from Africa, when she capsized and sank off the Welsh coast. There were no survivors.

The old days of passengers boarding a vessel out at sea came to an end with the opening of East India docks on the Thames in London in 1806 that allowed vessels to come alongside the quays for loading and unloading. But other changes were slower to come. Sail continued in use for trips to India and onwards to Australia long after change had come to other ocean voyages.

The East Indiaman *Euphrates* painted by Samuel Walters in 1835, just a year after she was launched in Liverpool. She is seen here off Cape Town. The ship was lost with no survivors in 1868.

A new route became important in the 1840s, following the discovery of gold in California. Clipper ships dominated the route from New York, round Cape Horn to San Francisco. It was always a long and arduous voyage. The record time was set by the clipper *Flying Cloud* in 1854 of 89 days 8 hours. It is a remarkable fact that the record remained in place right through to 1989. Even more remarkable is the fact that the navigation officer on this and other voyages was Eleanor Creesy. As a young woman, she had always dreamed of serving on a sailing ship and travelling the world, but recognised that the only way she could achieve her ambition was to marry a sea captain – and Josiah Creesy was that man. She learned navigation and became highly competent – as the record-breaking run clearly indicates.

Not all the vessels rounding Cape Horn made it back to port. Many of the more severely damaged ended up in the Falkland Islands, among them the *Snow Squall*. She was built in 1851 but in 1864 she ran aground off the South American coast when under way from New York to San Francisco. When she was brought to Port Stanley, she was found to be too badly damaged for repair, and much of her timber was used to construct a pier. Some, however, remained intact, and a salvage operation recovered the bow section and other remains. They are now on display in the Maine Maritime Museum, the only surviving remnants of the great American clippers.

A card advertising a clipper service from New York to San Francisco.

Chapter Two

The Start of the Steam Age

The first successful steam engines began work at the beginning of the eighteenth century. They were huge, cumbersome beam engines used for pumping water from mines. The pump rods, suspended from one end of an overhead beam, were raised by pulling down the opposite end of the beam via a chain attached to a piston in a cylinder. Steam was admitted below the piston, then condensed by spraying with cold water, creating a partial vacuum. Air pressure pushed the piston down. It was a single-acting system; the piston could pull, but couldn't push. In the 1760s, James Watt made the vital change that converted the old atmospheric engine into a genuine steam engine. He condensed the steam in a separate vessel, and instead of relying on air pressure, used steam power itself to move the piston. With this arrangement it was possible to allow the steam in both above and below the piston, moving it in either direction. It was no longer necessary to have pump rods to pull down one end of the beam now that the piston could push both up and down. Then, if instead of pump rods at the end of the beam you had a sweep arm and a crank, you could use the engine to turn a wheel. In fact, it was not really necessary to have the sweep arm at all; all you needed was a connecting rod to the piston and the crank. The engine could turn the wheels of industry – or a paddle wheel on a boat. The idea is really very obvious. If a water wheel could be used to turn machinery, then machinery could be used to turn a wheel in water. The water would be pushed backwards, and as Newtonian physics explained, every action has an equal and opposite reaction, so the vessel would move forwards.

One might have expected a British engineer to be the first to apply steam power to a boat, but instead it was a French aristocrat, the Marquis Jouffroy d'Abbans, who tried out a paddle steamer, the *Pyroscaphe*, on the Saône near Lyon in 1783. Any hopes of developing the idea were brought to a halt when revolution overwhelmed France. The next attempt to use steam power to move a vessel took a different route, with an American experiment to use a steam pump to draw in water at the bows and pump it out at the stern which in theory should have sent the boat on its way, but it was an idea not matched by available technology. It was then the turn of the British. William Symington built a small paddle steamer that he tried out on a lake near Dumfries in 1788. Among the passengers was the poet Robert Burns, but sadly he does not seem to have left an account of his experience. There was, however, a practical outcome as Symington was commissioned to design a paddle steamer for use as a tug on the Forth & Clyde Canal. The mechanism was simple, with a connecting rod from the engine to a crank to turn the single paddle wheel set in the stern. It was a success on its trials in March 1802, when it towed two 70-ton barges for twenty miles up the canal against a strong head wind in about six hours. But the company was worried that the water churned up by the paddles would erode the banks and the vessel was never put to use. The Duke of Bridgewater, however, was impressed and ordered similar vessels for use on his canal. Among the Duke's friends was the American engineer Robert Fulton, who began developing his own versions of steamers.

Fulton returned to America in 1807 where with Robert R. Livingston he built the paddle steamer *Clermont* to carry passengers up and down the Hudson between Albany and New York. It had two short masts carrying square sails and a 24 horse power engine, driving two 15-foot diameter paddle wheels set amidships. It was America's first successful passenger steamer. Britain was to get its own first passenger steamer in 1812, when a hotel owner, Henry Bell, hearing of Fulton's success, decided he could run a profitable service on the Clyde from Helensburgh. The *Comet* was, like the *Clermont*, designed with side

Robert Fulton's paddle steamer *Clermont,* the world's first passenger steamer that operated between New York and Albany.

paddles and carried a single sail that used the tall funnel as a mast. The success of these two enterprises encouraged ship builders on both sides of the Atlantic to start constructing paddle steamers. Mostly they were used on rivers or on short coastal trips.

There was, however, another obvious advantage in having steam power. It could be used to keep sailing ships going when the wind was in the wrong direction or simply not blowing at all. In 1819, the *Savannah*, that was being built as a conventional three-masted sailing packet, was bought by Captain Moses Rogers with the backing of the Savannah Steamship Company. He installed a 90hp steam engine to drive a pair of paddle wheels with the intention of using the vessel on the Atlantic crossing. He did not expect to use the engine for the whole voyage, but only when necessary. There was not enough space for fuel for the whole voyage and the paddle wheels were specially designed so that they could be lifted out of the water, folded up like a fan and stowed on board when not in use. She set sail from Charleston on 14 April 1819. En route she was spotted in the distance by the captain of the schooner *Contract*, who 'seeing the billowing smoke' assumed the ship was on fire and sailed to the rescue only to be astonished to find a steamer in mid Atlantic. Her appearance off the coast of Ireland caused considerable consternation and once again it was assumed that the vessel was on fire and a revenue cutter was sent to the rescue. Once aboard, the rather embarrassed revenue officers were given a tour of the ship by the captain. She arrived safely in Liverpool on 20 June.

The PS *Comet*, Britain's first sea-going passenger steamer, that operated on the west coast of Scotland.

It was, in its way, a triumph, but a hollow one. She had been gracefully fitted out, with two saloons, one for the men and one for the ladies, and each was panelled, carpeted with oriental rugs and hung with mirrors. The thirty-two state rooms were as elegant as any to be found on any ship of the day. All had been made ready for the passengers, but none had appeared. Steam boats were still regarded with suspicion, and even those prepared to travel on a river boat gave way to qualms when it came to crossing the Atlantic. As a result, the epic voyage was undertaken under ballast. In spite of the fact that steam had only been used for part of the voyage, the bunkers were almost empty by the time she arrived at Liverpool. Nevertheless, she excited considerable interest, and remained in port for nearly a month, before setting off for Stockholm, where King Charles IV was taken on a short cruise. There seemed a possibility for a time that the king might buy the vessel, but that came to nothing. The next stop was Russia, where the Tsar Alexander I also showed interest in buying the unique steamship, but again negotiations fizzled out without a deal. On returning to America, the *Savannah* was eventually used as a packet, but without the engines and paddles. Originally designed as a sailing ship, that was how she ended her days.

There was no great drive for using steam on the Atlantic, which is hardly surprising given the lack of enthusiasm shown by the travelling public. It made little economic sense to spend extra money on installing the machinery and allowing precious space on board to be taken up by the engine and bunkers for the coal, all for a possible saving of perhaps a day or two on the voyage. There was a general belief that using steam for the entire voyage was impractical, simply because not enough fuel could be carried. One possibility would seem to be to build bigger vessels, but most people thought that was not the answer. Dr Dionysus Lardner, an eminent British scientist, but one prone to sometime ludicrous errors, 'proved' that the whole thing was impossible. If you double the size of the ship, he argued, then you will need twice as much fuel, so nothing would be gained. This was another fundamental mistake. A ship's weight is supported by the water, a fact first established centuries ago by Archimedes, so the effort is needed to overcome water resistance, and to a far less extent air resistance. Water resistance depends on the area of hull submerged, and that is two-dimensional. But the volume of the ship is a cubic figure not a square one. Doubling the size does not double the amount of power needed to move the ship, and therefore does not require twice as much fuel. Bigger, in this case, is definitely best. Fortunately, there were engineers around more astute than the learned doctor, and one of those was Isambard Kingdom Brunel.

Brunel had been appointed chief engineer for the Great Western Railway, but one doubting Thomas on the company's board suggested that it might be impossible to build a railway that ran all the way from London to Bristol. With typical bravado, Brunel famously replied that he saw no reason to stop there – why not go on to New York? One other member of the board, Thomas Guppy, took the suggestion seriously. Probably Brunel himself had not even been that serious; when someone challenged one of his more extravagant ideas, he usually came up with one even more extravagant. But now the whole concept of a steamship to cross the Atlantic excited him. A consortium was formed of Guppy, Captain Christopher Claxton, a former naval officer and then in charge of the Bristol docks, and a third GWR visionary, Peter Maze. Together they set up the Great Western Steamship Company. Brunel was to design the vessel and it was to be built at the Bristol shipyard of William Patterson.

The ship was to make use of the latest technology in ship building. Robert Seepings had introduced new ways for strengthening the hull of a wooden ship. He had used diagonal iron straps and replaced the old wooden knees that joined the deck planks to the hull by iron knees. This prevented hogging, the tendency for large wooden vessels to droop at bow and stern which in extreme cases could lead to the ship breaking its back. The timber was protected against rot by Kyanisation, a system developed and patented by John Kyan, using mercury dichloride. The bottom of the hull was protected from attack from the marine worm by copper sheathing. She was to be an unusually large vessel for the time at 1,340 tons and would need a correspondingly big engine. The contract went to Maudslay, Sons & Field, a company well known to the Brunels. The founder, Henry Maudslay, had constructed the machinery for Isambard's father Marc's block making mill and had provided steam engines for use during the construction of the Thames tunnel.

The engine was a side lever, rather like a beam engine except that there are a pair of beams below the top of the cylinder, one end being connected to a cross head above the cylinder, the other to the cranked axle of the paddle wheels. This type was common in steamers as it had a low centre of gravity, adding to stability. There were two cylinders, each able to work independently. They were big engines and needed to be as they worked at a niggardly pressure of just 5 pounds per square inch (psi) but produced 250 indicated horse power. In order to feed the boilers with fresh water, as opposed to corrosive sea water, she was also fitted with the newly invented Hall condenser, which condensed the steam exhaust into pure water. She also carried sail on four masts.

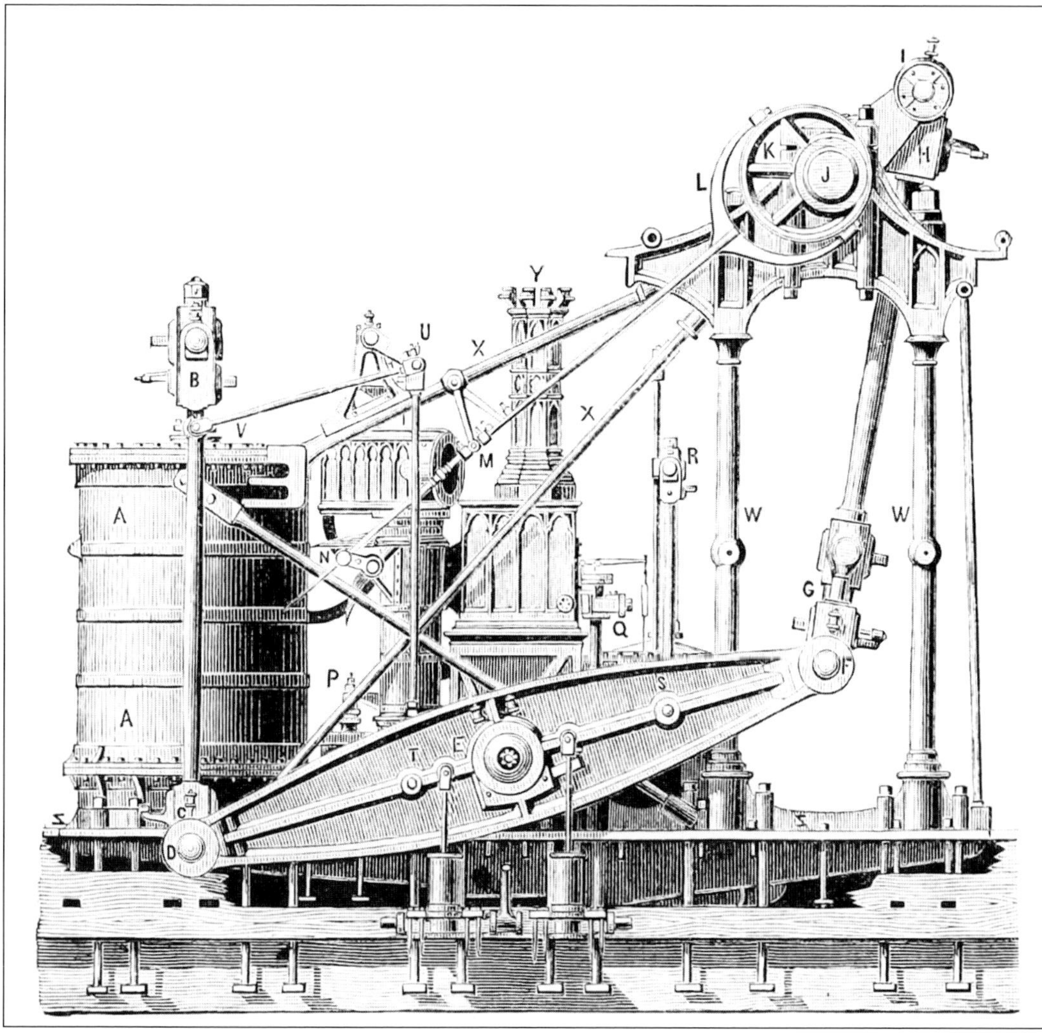

A side lever steam engine of the type regularly installed in paddle steamers.

As construction got under way, another company also began plans for an Atlantic crossing under steam. Junius Smith was an American who had set up in business as an exporter in England. When he took his family to visit America, he found the sailing packet that they travelled in was lamentably slow, taking 44 days for the voyage. He became an enthusiastic advocate for steam. Having failed to interest the British government, he decided to set up a private company, with the initial aim of building a fleet of four vessels. He began an association with the Birkenhead ship builder MacGregor Laird and they set up the British and American Steam Navigation Company. In 1836 they put in an order with the Carling and Young shipyard in London to build the hull of a vessel originally named *Royal Victoria*, later changed to

British Queen. Work was well under way when bad news arrived from Scotland – the company making the engine had gone bankrupt. There would be inevitable delays as a new engine was ordered from Napier.

Meanwhile things were going smoothly at Bristol and at the launch ceremony on 19 July 1837 she was officially named *Great Western* and floated out into the Cumberland basin in front of a crowd of some 50,000 spectators. Accompanied by a tug and a sailing packet, she now made her way round the coast to the Thames, where she was to be fitted out with her engine and other machinery. In London, the vessel attracted a huge amount of interest. It was not just her size, but the luxury on board that impressed. The star attraction was the grand saloon, 75ft long and 22ft wide, decorated with fifty large painted panels, covering everything from romantic landscapes to scenes representing the arts and sciences, and a series of smaller panels showing figures from mythology. The overall style was fashionable gothic and one commentator declared that it was a room that could 'vie with those of the clubhouses of London in luxury and magnificence'. It was estimated that 1,000 visitors came to see and marvel, including the Duke of Wellington and the Lord Mayor of London. In order to keep the passengers happy, Brunel invented a bell mechanism that could be used to summon a steward. The passenger pulled a bell rope that set a bell in the steward's room tinkling. At the same time, it activated a tin tab that popped up with the cabin number through a slit in a box.

The PS *Great Western* being towed by a steam tug out of Cumberland Basin, Bristol, where she was built and on her way to London for fitting out.

The rival company must have watched all this with some dismay, and they set about finding a way in which they could claim the honour of the first steam crossing. With their new ship too far behind schedule, they began looking for a suitable vessel and found one – a coastal steamer belonging to the St. George Steam Packet Company, the *Sirius*. She was a compact two-masted vessel of just 700 tons, roughly half that of the *Great Western*, and her previous voyages had extended no further than from London to Cork via Plymouth. Now she was to be got ready for the Atlantic. She had space for just 40 passengers, while the Brunel ship had '128 sleeping places for one class of passenger … and good places for servants'.

The two vessels were in London at the same time, and the *Great Western* was going through its first sea trial on 28 March 1838, when *Sirius* steamed past them, not on trial, but heading for Cork to take on fuel for the Atlantic crossing. On 31 March, Brunel was on board when his great ship set off for Bristol with, everyone was sure, quite enough time to overtake the rival. They were still in the Thames estuary when smoke billowed up from the forward boiler room. The chief engineer, George Pearne, at once made his way through the smoke and opened up the boiler feed cocks and put on a feed plunger. It was soon apparent that it was the boiler lagging that had overheated and caught fire, burning some of the deck planks and beams. Captain Claxton rushed to help Pearne, and Brunel himself decided to see what was happening down below. He had just started down the ladder into the boiler room, when a rung gave way and he fell 18 feet in what would have been a fatal accident if he had not landed on top of Claxton. Even so, the engineer was injured and unconscious and he had to be hauled back to safety.

The captain on board, James Hopkins, decided to ground the vessel to assess the damage and to get the injured Brunel ashore. He headed for Canvey Island and the vessel was grounded on the Chapman sands. A boat came alongside to pick up Brunel, who was to be bed-ridden for many weeks, recovering from his plunge into the bowels of the ship. Meanwhile the voyage got back under way at the next high tide. The ship arrived at Bristol on 2 April, where rumours had spread that Brunel was dead and the vessel was lost. The rumours did nothing to encourage the passengers who had booked for the voyage, and many cancelled, leaving just seven stout souls to take part in the epic adventure. Everything was ready to sail on the 7th, but their rival had left on the 4th, and Cork was a whole day's sailing nearer to New York. The race seemed lost before it had begun.

At first, *Sirius* enjoyed a calm passage, but eight days out to sea she had to hove to in order to refasten some of the boards on the paddle

The *Sirius*, the small ship that took the honour of being the first to cross the Atlantic using steam for the whole voyage.

wheels that had come loose. Four days later, they ran into a heavy storm, but Captain Roberts was happy to note that the vessel 'was behaving nobly and rode like a duck'. On 17 April they hit an even worse storm, which even the apparently unflappable captain described as dangerous and was accompanied by 'snow to cool our energies'. The little vessel ploughed on and on 22 April she finally reached land and signalled for the pilot to bring her into New York harbour, but no one appeared. The captain had no choice but to put her bows onto a mudbank, having already stopped the engines. Next morning, she floated free and made her triumphant entry into New York, where she was welcomed by a huge crowd. To Captain Roberts and his crew went the honour of bringing the first ship across the Atlantic using steam the whole way, and the extent of their achievement cannot be overestimated. This was a vessel designed for short crossings, mostly within sight of land, not for extended ocean voyages. Meanwhile, the rival was also nearing shore.

The *Great Western* hit the same storms that had plagued *Sirius* and those passengers who had not cancelled must have wished they had joined the rest as one by one they were laid low by seasickness.

Then their nights were disturbed by a terrible clanking – 'a dance of pot-hooks and frying pans was nothing to the din of the glorious clatter among the moveables that accompanied it'. Most of the row came from bolts that had worked loose on the paddle wheels. If the passengers were unhappy, the crew below decks were not much better off. They had the task of keeping the fires burning in the boilers, constantly shovelling in the fuel and moving it from the bunkers, while the ship rolled and tossed. Engineer Pearne persuaded them to keep on working by offering them extra pay. Everyone agreed except one stoker called Crooks, who left his post and proceeded to get drunk. The captain ordered him to be 'secured to the poop', but Crooks resisted, went for the captain and tried to push him overboard. He was eventually overcome and would have been clapped in irons for the rest of the journey if his shipmates had not insisted on his release, threatening to stop work themselves. So, Crooks was led back to his bunk, and the captain brought him medicine to cure his hangover – one dreads to think what dreadful concoction that might have been that was prepared for the man who had tried to drown him.

As *Sirius* was docking in New York, the *Great Western* dropped anchor off Sandy Hook, a spit of land that marks the furthest extent of the New Jersey shore on the approach to New York. Then, as she steamed past Ellis Island, she was awarded a 26-gun salute – one for what was then each state in the Union. Like *Sirius,* she received an enthusiastic reception and the New York papers published fulsome accounts of the vessel – 'she looked black and blackguard, rakish, cool, reckless and forbidding' as James Gordon Bennett wrote in the *Morning Herald.* The Brunel ship still had 200 tons of coal in her bunkers, where *Sirius* had virtually used her entire stock of fuel. It was even rumoured that cabin doors had been taken off and burned to keep her going at the end – which was certainly not the case. But what was very clear was that the *Great Western* had proved its superiority, not just in terms of fuel, but from the fact that she had been gaining on her rival, travelling on average two knots faster. Big had now actually been proved to be best. It was clear that *Sirius* was not really up to making regular voyages across the Atlantic, and was soon back on her old route between England and Ireland, only to be wrecked in 1847 off the Irish coast with the loss of twenty lives.

The British and American Steam Navigation Company soon had their own ships in service, ones that unlike *Sirius* were always intended for this route. The first into service was the *Royal William* and she made a crossing from Liverpool in 19 days, the first to use the port on the Mersey that was to become a regular starting point for ships to America. She was followed by the *British Queen* and the

Liverpool, but none were able to match the Brunel ship's performance. The *Great Western* continued in regular service and by 1844 had made 54 crossings with an average of just over 15 days in each direction. Where, due to the direction of the prevailing winds, the difference between eastern and western voyages in the sailing packets were seventeen days, the difference here, where the effect of wind was far less, was a mere three hours. No one could now doubt where the future lay – with large steamships, and the steamer was about to undergo another transformation, once again thanks to Brunel.

The arrival of paddle steamers on the Atlantic did not immediately end the role of the sailing packets, and for many the new steamers were interlopers, ruining a proud tradition going back centuries. John Malcolm Brinnin, in his book *The Sway of the Grand Saloon*, quotes an anonymous New Yorker, who was expressing the view of many others.

> In light head winds and moderate weather, a steamer would go wheezing and bubbling alongside of the proudest ship in the British or American navy, and passing laugh it to scorn: but let the ocean be lashed, and let the waves run high as the topmast, and how is this long stiff vessel, overburdened with the weight of machinery, with a burning volcano in her bowels, to ride on the crested billows and sink again into their dark, deep caverns?

In 1838, the Great Western Steamship Company announced that timber had been ordered for a new vessel that would be the same size as the *Great Western* and would be called *City of New York*. Within two months, Brunel had changed the plans. The size was to be increased: the length from 205ft to 254ft, the breadth 35ft 4in to 40ft and only the depth of hold remained constant at around 23ft. He must, however, have had doubts about the viability of a wooden hull at this size. He was right to be concerned, as was proved by a tragedy at sea in 1840.

Junius Smith had ordered a new steamer that was to be far bigger than the *British Queen*. The *President* was actually thirty feet longer than the *Great Western*, then the largest ship afloat. Naval architects had shied away from constructing a wooden vessel of this length, because of the danger of hogging. The new ship made a less than impressive start. Her maiden voyage to New York in August 1840 took more than sixteen days, and on the return journey to Liverpool in November of that year, she hit such bad weather that she was slowed to the point where there was not enough coal in the bunkers to make port and the Captain had no alternative but to return to New York. Then in March 1841, she set sail again for Liverpool, alongside a sailing packet, the *Orpheus*. They were hit by an immense gale that ripped the sails of the latter to shreds and left the *President* battling huge waves. In the height of the storm, the captain of the sailing packet lost sight of the steamer – and she would never be seen again. Eventually, wreckage was found, washed ashore on the Cape Verde Islands. What exactly happened will never be known, but there is a strong probability that, as experts had predicted, a wooden vessel of that length was fundamentally flawed. The wreck did not directly affect Brunel's thoughts – but he had listened to those who advised that there were limits to the size to which a wooden ship could be built. An alternative was available. He began to investigate the possibility of building in iron.

The iron ship was a true child of the industrial revolution. Iron manufacture had moved from the old charcoal furnaces to new ones using coke as a fuel. These, however, produced cast iron, ideal for construction, being very strong in compression, but of no use for ship building. Fortunately, Henry Cort had developed a system for converting cast iron into wrought iron, which when heated could be squeezed between rollers to create metal sheets. Three years later, the Shropshire iron master John Wilkinson built an iron barge for use on the Severn, called, appropriately, *The Trial*. After its launch he wrote: 'It answers all my expectations, and it has convinced the unbelievers, who were 999 in a thousand.' But a river barge is a long way from an ocean-going ship. A number of small iron vessels were built over the next few years, but it was only in 1822 that one went to sea. Charles Manby took out a patent for iron steamboats and began manufacturing at his works at Tipton at the heart of the Birmingham Canal system. The first vessel, the *Aaron Manby,* named after his father, was ordered for France but was too big to travel down the English canal system. So it was sent in pieces to London, reassembled and set off down the Thames and out across the Channel to Le Havre.

After that, progress was slow but steady. William Laird had established a ship yard at Birkenhead on the Mersey and decided that the future lay with iron ships. His first vessel was a lighter for use in Ireland, but he was determined to be more adventurous. There was, however, a problem that needed to be overcome. Navigation depends on the magnetic compass, and the needle is attracted to iron. This, it was thought, made sea going iron ships impossible to use out of sight of land. It was an astronomer, Professor Airey, who pointed out that the magnetism of the ship's iron was a constant. It could be corrected by positioning counterbalancing magnets alongside the compass. Brunel and his associates were already discussing the use of iron, when the Laird packet *Rainbow* steamed into Bristol. It was the perfect opportunity to see how the vessel performed and Claxton made a number of voyages that convinced him that iron was far superior to timber for construction. The *City of New York* would have an iron hull and was given a new name – it was to be the PS *Great Britain*.

Construction of the iron ship was totally different from that of the traditional wooden vessel. Instead of creating a skeleton consisting of the keel, ribs and beams to which the planks were attached, the *Great Britain* had iron girders running from bow to stern, to which longitudinal and transverse bulkheads were attached, with decking above. This created a set of rigid boxes. The hull itself, however, looked back to the days of Nelson's navy in its overall shape. It had a pronounced tumblehome, broadening out above the waterline, then nipping back in towards the upper deck. The outer shell was made of comparatively small iron plates – the largest available at the time – riveted together.

The next big decision to be taken was the design and manufacture of the engine. The obvious choice was to return to the company who had made the engines for the *Great Western,* but their quote was very high and an alternative bid had come in from a young designer, Francis Humphrey, who had patented a trunk engine, being manufactured by Halls of Dartford. Essentially this was a space saving device. Instead of the usual arrangement of piston rod, connecting road and crankshaft, the piston itself was extended as a hollow cylinder or trunk. Brunel was dubious, but the company were attracted by the price and the order was put through. Then a snag arose. Halls were not prepared to manufacture the engine; they were not convinced that big steamers were going to be ordered in the future, and they were not prepared to install new machinery to make what could be a one-off engine. They suggested that the Great Western Company might like to take on the manufacturing work themselves. It was hardly a vote of confidence in the Humphrey design, but the company decided that they would indeed set up their own manufacturing company. Then a new problem appeared.

The engine called for a massive crankshaft to be forged. At the time this would have had to be done using water-powered tilt hammers, in which the hammer head was raised at an angle before falling back under gravity. In effect, it was a mechanical equivalent of the blacksmith and the anvil. But no one was able to accommodate such a large job. Humphrey, however, consulted the manufacturer James Nasmyth to see if he had a solution. In Nasmyth's own words, the letter 'set me a-thinking'. Instead of a water wheel raising a hammer by a small degree, what if he used steam power to lift it vertically? He sketched the idea out there and then and the Nasmyth steam hammer was to become an essential fixture in shipyards throughout the land. It was not, however, destined to be used to forge Humphrey's crankshaft.

From the start, it was assumed that the *Great Britain* would be a paddle steamer, but an alternative was being developed by a gentleman farmer called Francis Pettit Smith. His idea was based on an ancient device, the Archimedean screw. This consisted of a spiral screw set in a cylinder. If one end of the cylinder was immersed in water, and the screw turned by a handle or mechanism above the surface, water would be lifted up the tube. It was a device that had been used in Holland where, powered by windmills, it had been used for land drainage. Pettit Smith reasoned that if the whole screw was immersed and attached to a vessel, it would push water away from the boat and the boat itself would move forward. It would be a screw propeller. After experiments with models on the duck pond at his farm, he moved on to try his propeller on a launch, and then when he was satisfied, he installed the machinery in a three-masted schooner, *Archimedes,* and sent it out to advertise his invention. Brunel and his colleagues were intrigued, and Guppy joined the vessel to see how it performed. At the end of the period, he reported back favourably, and it was agreed. The paddle steamer plan would be abandoned – the ship would be a screw steamer, the *SS Great Britain.* Poor Humphrey! He had just solved the problems that stood in the way of building his engine, and now all was lost. Nasmyth described the result. 'Perhaps the disappointment of his hopes proved too much for him: and a brain fever carried him off after a few days' illness.'

The ship was finally launched on 19 July 1843 from the Patterson yard in Bristol. There was a great crowd for the event, including Prince

The SS *Great Britain* being fitted out in Bristol: this photograph by Fox Talbot is believed to be the first ever taken of any ship.

Albert, as the ship was hauled out of the building lock. Unfortunately, one of the ropes broke just as Mrs. Miles, who had the honour of christening the vessel, was throwing the bottle of champagne and it missed. Prince Albert promptly picked up another bottle and his aim was good. Then there was a problem getting her out of the Cumberland Basin and the entrance lock had to be enlarged. The sea trials went well, but when she left on her first voyage to London, she had got no further up the Bristol Channel than Lundy when she was hit by an immense storm. She weathered that until two waves hit simultaneously bringing the ship to a temporary standstill, before getting smoothly under way again. There was little damage apart from the loss of part of the figurehead. After that she made good progress with a maximum speed recorded at 14 knots.

In London, she aroused as much interest as her predecessor had done and was visited by the Queen and Prince Albert. One paper criticised the 'extravagance' of the carpeting, to which Claxton wrote an indignant reply: 'A large sum of money has not been uselessly squandered in procuring gaudy decoration. Its fittings are alike chaste and elegant.' This is certainly true of the saloon, where the only elaboration comes from the gilded capitals of the twelve columns that are lined up down the centre. Claxton also showed the true Victorian attitude when explaining the arrangement that had been made to preserve female modesty. They had rooms reserved for them with boudoir and sitting room: 'ladies who may be indisposed, or in negligee, will be enabled to reach their sleeping berths without their being the slightest necessity for their appearing in public.'

While the ship was performing well, the navigation officers were proving less satisfactory. On the first trip to New York, the vessel missed the opening that would take them into harbour, finishing up 35 miles to the north near Martha's Vineyard. A pilot brought them safely in. A far more serious error occurred in 1846. She set sail from Liverpool on a route that would have taken her past the southern tip of the Isle of Man and then she should then have turned north to round the Irish coast. For some reason the captain seems not to have realised that they had passed the island and simply ploughed straight on, until the vessel grounded in Dundrum Bay, County Down. Word reached Brunel that the great ship was irretrievably lost, and he rushed across to Ireland to see for himself. He found that nothing had been done to protect her from the elements. 'The ship is lying like a useless saucepan kicking about on the most exposed shore you can imagine with no more effort or skill applied to protect the property than the said saucepan would have received on the beach at Brighton.' He set about a programme of protection until such time as it was possible to

refloat her. In August, the effort was made and she was towed back to Liverpool for repairs.

When the estimates for the work came in, the cost was estimated at £22,000 but amazingly, she had only been insured for a measly £17,000 and the company had run out of funds. The vessel that had cost £120,000 to build was eventually sold to Gibbs, Bright & Co. for just £18,000. She remained in service, travelling the world, until in 1886 she was badly damaged in a storm while rounding Cape Horn, but made it

The *Great Britain* has been fully restored in her home port of Bristol; this is the main dining saloon on board.

to the safety of the Falkland Islands, where she was considered too far gone for repair and was left to rot in Sparrow Cove. That is, of course, not the end of the story. She was brought back to Bristol and the dock in which she was built, restored and now can be seen in all her splendour.

This has been rather a long description of just two vessels, but it is impossible to overstate their importance. The *Great Western* had proved that, given a big enough ship, Atlantic crossings by steamers could be profitable. The *Great Britain* was revolutionary, bigger than any ship yet built, with an iron hull and a propeller. One obvious advantage of the screw over the paddle wheel appeared in rough weather with a beam sea causing the vessel to roll. On the paddle steamer that would have resulted in one paddle being lifted clear of the sea to thrash uselessly in the open air, while the screw would remain submerged. Even so, there were those who doubted if the screw really was an improvement. That was settled in dramatic fashion in 1845, when two identical frigates were fastened together stern to stern. *Rattler* was screw propelled, *Alecto* had paddles. The order 'Full ahead' was given, and *Alecto*, her paddles thrashing the water, was towed backwards at 3 knots. It was now clear; from now on, Transatlantic steamers would have metal hulls and propellers and that would set the pattern for another century.

Chapter Three

The American Challenge

American sailing packets may have been the finest in the world, but the British completely dominated the world of steam. This was not a situation that pleased patriotic Americans. No one spoke more forcefully on the subject than the American Chargé d'Affaires to the Hague, Christopher Hughes. He had crossed the Atlantic on the *Great Western* and though he seems to have found no fault with the ship, he was deeply offended by the class-ridden attitude of the British. He compared that unfavourably with the true Yankee who 'never thinks of rank at all' and 'regards himself to be the equal of all – superior to none'. He added, rather more controversially – 'I believe that every American aboard this ship (even the coloured servants) agreed with him'. He declared himself to be a Yankee – and it is doubtful whether many of the white population of the southern slave states would have thought the views of 'coloured servants' as being worth considering at all. He concluded:

> Nothing but necessity induces Americans to use these English steamers. And we generally feel humiliated at the fact that this *monopoly* has been allowed to fall into English hands. The fact has lowered us in the opinion of foreign nations, as a nautical and enterprising people. *We ought to build steamers.*

He was not alone in believing that the British monopoly of Atlantic steam was an affront to the whole nation. In 1839, the Canadian Samuel Cunard had won a contract for his newly formed British and North American Royal Mail Steam Packet Company to carry mail between Halifax, Nova Scotia, and Liverpool. The Cunard ships were, however, not just carrying mail between Canada and Britain, but also were calling in at Boston for American mail. This was an affront to national pride. One incentive for Americans to build steam would be if the government was to offer a guaranteed payment for carrying the mail. Congress recognised this as necessary, and a Bill was passed in 1845 offering a mail subsidy to any company that would agree to build four steamers that would carry mail between America and any English port. The vessels would have to be at least 2,000 tons and they had to be built in such a way that they could, if necessary, be easily converted into warships. It was also expected that the ships would prove in every way superior to their British rivals. Among the enthusiastic supporters of the scheme was Junius Smith who, in spite of the loss of the *President,* was hopeful of getting the contract. He was to be disappointed and retired from the business altogether. There were two other contenders, one of whom, Edward Knight Collins, might well have won the contract, but his bid came in too late to be considered. So instead, it went to a New York businessman, Edward Mills, even though he appeared to have no previous ship building experience. He offered to provide a mail service to a British port and Le Havre, in return for which he had proposed a subsidy of $300,000 per annum. The government was, however, so delighted at the prospect of getting an all-American world-beating steamship line that they upped the payment to $400,000 for a ten-year period.

Governmental delight was short lived. Mills was unable to find the money to build just one steamer, let alone four. Time rolled on and nothing happened, until the Mills contract was bought out by a new group, the Ocean Steam Navigation Company of New York. By now, however, the government was regretting its rush of optimistic largesse, and cut the subsidy by half. In spite of this, the new company put in an order for a steamer with the New York company of Westervelt and Mackey. The new vessel was launched with great ceremony on 30 January 1847 and given an appropriately patriotic name – *Washington*. She was not exactly representative of the best and finest that shipwrights could devise. Quite modest at just 260ft long, she looked like the man of war that Congress had specified she might need to become. The designers had even gone so far as to paint false gun ports down both sides of the hull. As well as her steam engine to drive the paddle wheels, she had a full suite of sails carried on three masts. She would, however, every patriotic American was convinced, show the British just what American expertise could achieve. Her first run was destined not for Liverpool, which up to then had been the main transatlantic port of call, but for Southampton, from where she could make her way to Le Havre. She was scheduled to leave New York at precisely the same time as the Cunard steamer *Britannia* was due to sail from Boston. No one doubted, she would leave the British toiling in her wake.

Even before she set sail, there was a problem. It proved difficult to find an experienced crew for the engine room and as a result there was only one officer in charge and a motley band of men recruited from river boats. Unfortunately, many of those who were at home on the calm waters of inland waterways fell to seasickness once the ship was clear of harbour. As *Britannia* puffed happily along, there were days

The Ocean Steam Navigation Company's ship, *Washington*. The black and white hull is typical of a warship of the period, but the gun ports are fake. From an engraving by Nathaniel Currier.

when on the American rival there were simply not enough stokers to keep up boiler pressure. Nominally, it should have been possible to produce steam at 30psi but it seems that often only half that was available. As a result, *Washington* arrived in Europe two days later than the British rival. *The Times* correspondent, who had seen the vessel approaching, was not impressed; the ship, he wrote, 'seemed rather to roll along than steam through the water'. There was, however, one positive outcome of the voyage; it helped establish Southampton as a major port for future Atlantic crossings.

She left Southampton for the port of Bremerhaven, near Bremen in North Germany, where she delivered around 3,000 letters to be distributed throughout the country. On her return to Southampton, she struggled against a head wind and took 46 hours for the short voyage. Worse was to come. When she left Southampton to return to New York, she had scarcely got further than the Isle of Wight when she had to turn round and come back again. She had taken on board high-quality anthracite coal, excellent for steaming because of its high calorific value. It turned out, however, to be a bit too good: the intense heat melted the fire bars in the furnaces. When *Washington* eventually made it back to America, no one could claim that she had been the triumph the Americans had hoped for. It would be left to a new line to try and restore national prestige.

The Cunard steamer *Britannia* pictured in full sail off the English south coast.

Unlike the unfortunate Mills, Edward Knight Collins came from a family with a fine nautical tradition. Both his father Israel Collins and uncle John Collins had been captains. He was born in 1802, but his mother died shortly afterwards and he became a ward of his Uncle John, who had settled at Cape Cod. While still in his teens he became an apprentice to a commission house in New York. This gave him the opportunity to make numerous voyages up and down the east coast. He served for a time on ships trading to Cuba and Florida. His father retired from his sailing career and established a shipping agency in New York, and the turning point in his career began when he hired a schooner to sail to Charleston, South Carolina, to collect a cargo of cotton for England in 1824. The following year he established a packet boat line, linking New York to Vera Cruz. The following year, Edward joined him and the firm became Israel G. Collins & Son. His father died in 1831 and the firm was then simply E.K. Collins.

The business was a roaring success and Collins used the profits to build a whole fleet of magnificent sailing packets, named after leading theatrical figures, such as Garrick and Sheridan, earning it the nickname 'the Dramatic Line'. They were acknowledged as being among the finest ships on the Atlantic run. But he was a man with an eye to the future and had been thinking of converting to steam as early as 1840, so it was a real setback when he lost the chance to get his bid accepted for the first mail steamers. Now, however, with the rather dismal failure of *Washington* to impress anyone, his moment had arrived.

Congress was still anxious to have a dependable mail service to Europe, carried in American ships. Senator Bayard of Delaware spoke for many patriotic colleagues when he declared that cost should not be the main concern, but what was needed was to give an American shipping company 'a completely free hand to proceed with the absolute conquest of this man Cunard'. Collins stepped in to offer to carry the mails between New York and Liverpool using five brand new steamers, making twenty round trips a year. In return, the government would pay him a subsidy of $375,000 per annum in quarterly payments. Congress dithered for over a year, before finally agreeing to sign a contract with Collins in November 1847. The new company was officially called The New York and Liverpool United States Mail Steamship Company, but for most people it was simply the Collins Line. He had financial support from two bankers and was now ready to start building his steamers.

The first two, the *Atlantic* and the *Pacific,* were built at the yard of William H. Brown in New York. *Atlantic* was the first to be launched in 1850. The wooden hull was reinforced with diagonal iron strapping and her vertical bow and lack of a bowsprit gave her a unique appearance. She was a large vessel, 284ft long by 46ft beam, and was powered by two side-lever engines built by the Novelty Iron works, also of New York. These were immense machines, each with a 95in diameter cylinder and a piston stroke of 9ft. To keep them going, the four boilers had to be fed with around 85 tons of coal a day. The paddle's wheels were 36ft in diameter. As well as her steam engines, she also carried sail and was rigged as a barque. Her interior was luxurious and will be described later, but there were a number of features that were unique at the time and designed to ensure that the passengers enjoyed luxuries seldom available on other craft. She was steam heated and had a vast ice room that could hold 40 tons of ice that could be used to keep food fresh for the voyage. She was launched to the acclaim of a huge crowd and the expectation was that she would prove the most powerful vessel afloat.

She set off on her maiden voyage on 27 April 1850 with a crew of 170 but only 100 passengers; the public might have praised the new ship but were not yet bold enough to risk travelling on her. She had scarcely cleared land before she ran into an ice field and although she ploughed through it, the paddle wheels were damaged in the process and her speed was reduced. On the sixth day, one of the condensers packed up, which was essential for providing feed water for the boilers.

She had to heave to, while repairs were carried out. As a result, instead of a record breaking voyage, she finally reached Liverpool after sixteen days at sea. It was not the triumph America had hoped for, and the British were understandably cock- a-hoop and not slow to criticise what they considered an ungainly vessel. American honour, however, was restored when, after all the necessary repairs had been made, she made her way back to New York in a record time of 19 days, 16 hours. By the time all four vessels were in service, the first two having been joined by the *Arctic* and the *Baltic*, they were regularly making fast crossings.

On 9 April 1852 the *Pacific* became the first vessel of any kind to cross the Atlantic in less than ten days, having arrived off Sandy Point from Liverpool in 9 days 20 hours 15 minutes. Even the British were impressed, and *Punch* celebrated the event in verse.

A steamer of the Collins line,
A Yankee Doodle notion,
Has also quickest cut the brine
Across the Atlantic Ocean.
And British agents, no way slow,
Her merits to discover,
Have been and bought her – just to tow
The Cunard packets over.

The Collins paddle steamer *Baltic*.

It is doubtful if Cunard appreciated the joke. He had responded with a new ship, *Asia,* built on the Clyde, that was fitted out with Napier engines, and was expected to cruise at 12½ knots. But she did not offer the comfort and luxury of the Collins ships, and the latter were now carrying almost two thirds of the passengers travelling the Atlantic by steamer, though the numbers were still far lower than those still taking the sailing packets.

The Collins ships may have been successful in winning passengers, but they were a good deal less successful when it came to the all-important matter of making money for their shareholders. One of the factors that made the vessels so popular was the luxury of the accommodation and the excellence of the meals served on board. The illustration of a breakfast menu gives a good idea of what was on offer – and dinners were even more sumptuous. But all this cost money. Cunard, by comparison, was considered quite frugal and the costs lower – a Cunard captain, for example, was paid $2,500 a year but the Collins men got $6,000. The end result was that, even with the subsidy, Collins was losing $17,000 a trip. He launched a campaign for more funds in 1852, sending his newest vessel, the *Baltic,* up the Potomac to Washington and entertained Senators and Congressmen on board. The lobbying succeeded, and his subsidy was raised from $19,250 per trip to $33,000. He was very much back in business.

The breakfast menu for the *Arctic;* no one had to go hungry.

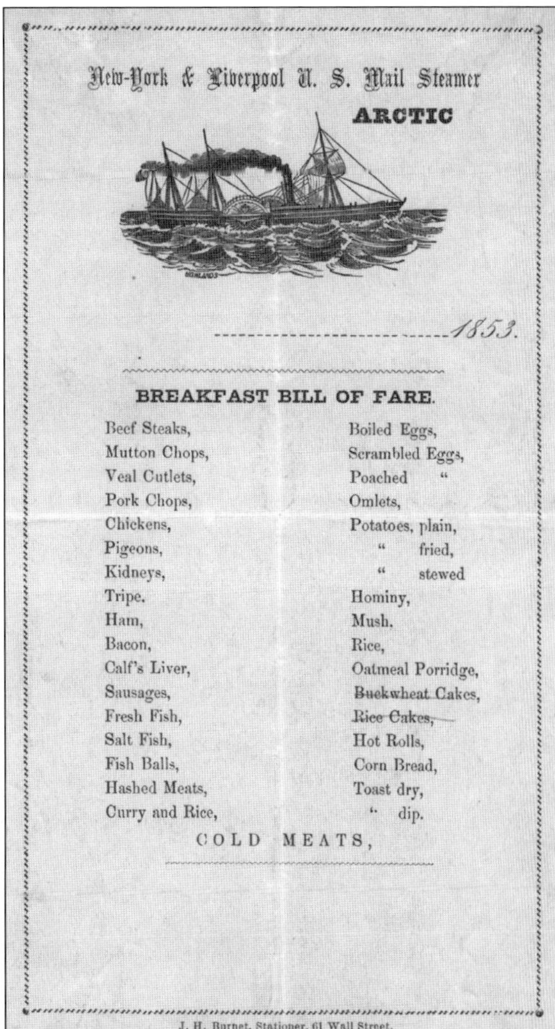

New-York & Liverpool U. S. Mail Steamer

ARCTIC

------------1853.

BREAKFAST BILL OF FARE.

Beef Steaks,	Boiled Eggs,
Mutton Chops,	Scrambled Eggs,
Veal Cutlets,	Poached "
Pork Chops,	Omlets,
Chickens,	Potatoes, plain,
Pigeons,	" fried,
Kidneys,	" stewed
Tripe.	Hominy,
Ham,	Mush.
Bacon,	Rice,
Calf's Liver,	Oatmeal Porridge,
Sausages,	Buckwheat Cakes,
Fresh Fish,	~~Rice Cakes,~~
Salt Fish,	Hot Rolls,
Fish Balls,	Corn Bread,
Hashed Meats,	Toast dry,
Curry and Rice,	" dip.

COLD MEATS,

J. H. Burnet, Stationer, 61 Wall Street.

All seemed to be going well for the Collins line until September 1854. The *Arctic* had left Liverpool but seven days out, on the 20th, she ran into fog on the Grand Banks off the coast of Newfoundland. In those days, long before the invention of radio let alone radar, the only way in which other shipping could be avoided in such conditions was by keeping a good watch and sounding a foghorn. It would have been prudent to slow down in such conditions, but there was pressure on Collins' skippers to make fast times and a general view that the faster you went in fog, the sooner you would be clear of it. So, *Arctic* ploughed steadily on until the look-out in the bows cried out a warning of collision with another steamer, the French screw steamer *Vesta,* a small ship used by a major fishing company to ferry fishermen between Newfoundland and Saint Pierre Island. The order was at once given to turn hard-a-starboard and cut engines. Captain Luce rushed up on deck, but by then the two vessels had collided. No one on the *Arctic* was even aware that anything serious had happened. One passenger in the saloon noted that there was 'a slight shock, although it was scarcely more than a tremor or a quiver'. At first, Luce thought his vessel had scarcely been harmed, but that the *Vesta* was probably fatally damaged, '[she] seemed to be literally crushed or cut off for full ten feet'. He ordered a lifeboat overboard in charge of his first officer Robert Gourlay to offer help to the crew and passengers of *Vesta* and a second lifeboat was then soon launched.

What Luce had not appreciated at first was that his ship had not been hit by another wooden vessel but one of the new generation of iron-hulled steamers and that she had forced extensive holes above and below the waterline and the pumps were set working flat out. Unlike more modern vessels, including *Vesta, Arctic* had no watertight compartment, just one long open hull. Attempts were made to cover

The wreck of the *Arctic* following her collision with the *Vesta.*

the holes with sail cloth, but water kept coming in. Luce decided that his only option was to abandon *Vesta* to her fate and make for land as quickly as possible. He was about four hours away from Cape Race on Newfoundland and he sent instructions to the engine room to go full speed ahead. He attempted to recall the two lifeboats but was unable to contact Gourlay and was forced to abandon him and his crew. They had scarcely got under way than they ran straight into one of the *Vesta*'s lifeboats; only one of the dozen on board survived, a fisherman, François Jassonet, who was hauled on board *Arctic.*

It soon became apparent that the pumps were quite unable to keep pace with the speed of the water flooding in, which eventually reached a level in the engine room where the fires were extinguished. Now, with only her sails to help her move, the ship floundered and Captain Luce ordered the remaining lifeboats to be lowered and that female and child passengers should be given priority. But, with the ship sinking, orders were flagrantly ignored. One boat was being prepared for lowering into the water, when it was mobbed by members of the crew, overturning it, sending the occupants tumbling into the sea, where all but three drowned. It was righted and in spite of the orders, it was again taken over by men, passengers and crew. On the opposite side of the ship, the chief engineer commandeered one lifeboat, claiming it was necessary for him to make a final attempt to plug the holes. With other engine room staff, the half full boat simply rowed away. As the ship began to sink towards an inevitable doom, there were still 300 left on board and just one remaining lifeboat. Luce was still desperate to save those he could and ordered timbers to be thrown overboard and the lifeboat lowered, so that a raft could be lashed together. But once again, the boat was overwhelmed and there was nothing to be done but to cut loose and row away, leaving the raft incomplete.

Luce made one desperate effort to get help. He instructed a young trainee engineer, Stewart Holland, to take a place in the bows with the single ship's cannon and fire it off at regular intervals in the hope of attracting any nearby vessel. Holland was still at his post when the vessel finally disappeared beneath the waves. Captain Luce stood with his son at his post to the last, before they were washed overboard. Luce later described the scene that faced him as he bobbed up to the surface. 'A most awful and heart-rending scene presented itself to my view – over two hundred men, women and children struggling together amidst pieces of wreck of every kind, calling on each other to help, and imploring God to assist them.' He had hardly had time to take in the scale of the catastrophe, when one of the paddle boxes rose to the surface, striking his son and killing him instantly. Luce himself was able to cling on to it.

Two of the ship's lifeboats made it safely to Greenland, where they discovered that the *Vesta* had not sunk after all, but thanks to her watertight bulkheads had stayed afloat and staggered into harbour. Two vessels were sent out to look for more survivors. A third lifeboat was picked up by a sailing ship and they also came across the makeshift raft, only to find that of the seventy-two who had clung to it for safety, only one, crew member Peter McCabe, had survived. Those who had clung to various bits of wreckage survived for two days before a sailing vessel 'the *Cambria*' spotted them, and a few more survivors, including Captain Luce, were taken on board. Those who had set out on the three other lifeboats were never found. Luce never went to sea again.

The contest between Collins and Cunard was temporarily halted as a result of the Crimean War, but when that ended in 1856 the rivalry resumed. Cunard had already ordered new ships from Napier, and the *Persia* was to be his most powerful vessel yet. It was a paddle steamer but was iron hulled and had a massive two-cylinder engine, with each cylinder being 100 inches in diameter. In trials she achieved a maximum speed of 17 knots but was expected to be kept to 13 knots to conserve fuel – as it was, she still consumed 145 tons of coal a day. In January 1856, Collins' *Pacific* sailed out of Liverpool, with her crew of 141 and just 45 passengers: confidence had been badly hit by the *Arctic* tragedy. *Persia* was still in Liverpool, but everyone was convinced that there was no need for any sense of urgency: she would easily overtake her rival. Five days out, she unexpectedly ran into an extensive ice field. She suffered considerable damage but, thanks to her iron hull, she withstood the impact and was able to make a steady progress towards New York, all thoughts of records and races forgotten. Everyone expected to see *Pacific* berthed in New York, but she had never arrived. The facts will never be known, but the likeliest explanation is that she ran into the same ice field that had almost crippled the Cunarder, but her wooden hull had not withstood the impact. Half the Collins fleet had been lost at sea.

Collins tried to retrieve his position, starting by trying to reassure potential passengers that, in spite of what had happened, the ships were safe. He set out adverts stressing that the ships would now have watertight bulkheads and that in future they would not cross the Banks after 1 August to keep clear of ice. A new ship was ordered, the *Adriatic*, which at 354ft long and 50ft beam, would be the largest vessel then afloat. She was intended to have all the latest technical developments available so that not only would she be the largest but also the most powerful ship on the Atlantic. Her two steam engines had immense 100 inch diameter cylinders, and were initially fitted with the patent valves designed by the American Horatio Allen. These were cut-off valves, which could vary the time during which steam entered the cylinder, so that when working the steam could be cut off before the piston had reached the end of its passage down the cylinder, and for the rest of the stroke, work would be done by the expanding steam, making for more efficient and economical working. The idea was sound, but in practice the valves proved unsatisfactory and the engines had to be refitted with conventional valves. There was also trouble with the condensers that overheated and they too had to be changed. All this delayed the completion of the vessel and although, like all the Collins vessels, she was grandly appointed, when she finally left New York in November 1857, she had 188 crew on board, but only one in ten berths were filled.

The final blow for Collins came when the government finally decided that they were paying Collins far too much in subsidy and cut it back to the original $385,000. The *Adriatic* was scheduled to sail from New York in February 1858, but a brief notice in the press announced that the sailing was cancelled. The Collins ships were all put up for sale and were bought for a miserable sum of just $50,000 by Dudley Fuller. The Collins line that had flown the American flag and competed with Cunard was no more.

The last of the Cunard paddle steamers, the *Persia*, from a painting by David Lyle.

Chapter Four

Building the Ships

The early paddle steamers all had wooden hulls, and the basic vessels were built exactly as if they were sailing ships. Indeed, some were, like the pioneering *Savannah*, simply sailing ships into which the machinery was later added. The technology of wooden ship construction had scarcely changed for centuries and one of the most innovative of the early shipwrights was Matthew Baker, who was appointed Master Shipwright, designing vessels for Queen Elizabeth I's navy. One of his innovations was to introduce a measure of a ship's size, its tonnage, or as it was originally known 'tunnage'. It was defined by the number of tuns of Bordeaux wine that could be fitted into the hold. Modern tonnage is, of course, different. Naval vessels are measured by the weight of water displaced by the ship – which is the same as its weight. Merchant vessels are defined by the weight of cargo they can carry, which is very different from Baker's measure, and is basically one of volume. Today, the ton has given way to the metric tonne.

Far more important was his method of laying out a ship's lines on paper. He set out cross sections at regular intervals along the hull, based on a variety of curves. These could then be translated into thin wooden patterns in the mould loft which could then be used to accurately shape the ribs to be installed at those points. It became a standard technique in ship building. But if this suggests that everything was now becoming quite scientific, this was not really the case. The master shipwright's designs were drawn, but the rules under which they were drawn were a closely guarded secret. One apprentice, who signed up to learn the craft, paid £3 for the privilege of being taught, but a clause in the contract informed him that he would be fined £10 if he tried to pass on his skills to others. There was a mystique, and the truth was that the shipwright had no real means of assessing how his design would work in practice, apart from past experience. Even as late as 1852, the *Nautical Mirror* carried this article on wooden shipbuilding:

> Science has comparatively little to do with the matter. A few general principles, no doubt, gave a basis, but the superstructure was greatly an affair of guess-work and eye-work. Frames were laid down from 5 to 10 feet apart, near the centre of the vessel. Others were placed

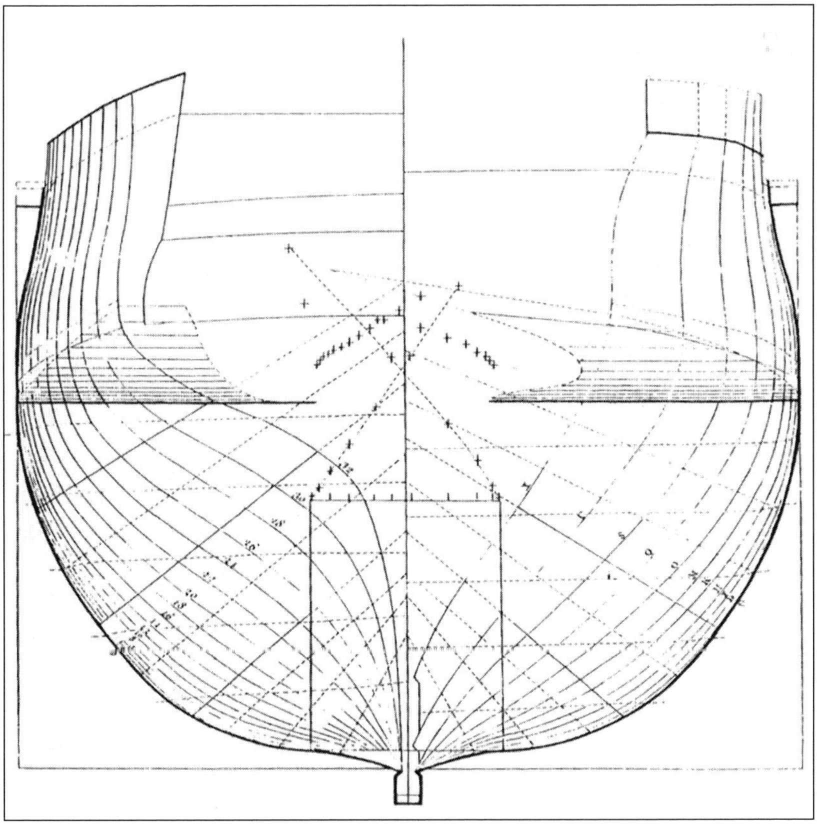

The body plan for a wooden ship, showing the lines from stern on the left and from the bows to the stern on the right: from Rees's *Cyclopedia* of 1819-20.

> at a distance of 20 feet or so, from the bow and stem; and all the rest of the work was done at the arbitrary pleasure of the builder, by his individual caprice or conjecture.

Earlier vessels of the Tudor period were clinker built, in which the outer planks of the hull overlapped, and it was this that gave the hull its main strength. By the time of the paddle steamers, however, they would have been carvel built, with the planking smoothly abutting instead of overlapping. Construction would start with the laying of the keel. This would generally be too long for a single timber, so pieces were joined by scarf joints, with the ends shaped to overlap, so that seen in profile, the joint would appear in the shape of a letter 'S'. Above that was the keelson, then the ribs would be added, together with the stern and bow posts. Once they were in place, the outer planking could be added. Originally this would have been sawn by hand at the saw pit, where the top man pulled the saw up, while his partner down in the pit, pulled it down. The top man had the harder task, but the other spent his working days covered in sawdust. By the early nineteenth century, mechanical saws, powered by steam, were in use in the bigger shipyards. Curved timbers were carved with a special tool, the adze, like a conventional axe but with the blade turned through 90 degrees.

In order to fit the straight planks to a curved hull they had to be steamed, which made them pliable. While still hot, they were clamped to the ribs and then nailed in place. The ribs were joined to the sides by beams, held in place by knees – rather like very large shelf brackets – and deck planking added. Once the hull was complete, it was made watertight by caulking, stuffing material, such as old pieces of teased out rope into the cracks, then pushing them home with a tool, rather like a blunt chisel. The seams were then covered with tar.

All the early steamers carried sail as well as using steam, so masts and rigging had to be supplied. The rigging required a great deal of rope – it has been estimated that Nelson's famous ship *Victory* would have had around 26 miles of rigging altogether, made mostly from hemp. As with the sawyers, rope making was mechanised by the end of the eighteenth century. A classic example survives at Chatham, first used in 1811, with machinery designed by Henry Maudslay, some of which still remains in use. Its size was determined by the length of cable required for use with an anchor in water at a maximum depth of 40 fathoms (240ft). In practice, that meant the works would have to produce rope at 120 fathoms, so the building had to be a quarter of a mile long. Individual threads are secured, then passed to hooks fastened to a plate on a wheeled carriage at one end of the rope walk. The carriage is then pulled back and as it moves, gearing rotates the plate, twisting the strands together to make rope. The process then has to be repeated with rope to make thicker cables.

Shipbuilders on a wooden vessel at the David Williams yard, Porthmadog, in the early 1900s. The man on the left holds a caulking hammer and the one next to him an adze.

The rigging was controlled through blocks, which had been made by individual craftsmen, each going through all the processes to make the complete block, which is basically a pulley inside a curved wooden frame. The process was mechanised by Marc Brunel, who broke the construction down into its constituent parts, and devised specific machines for each operation. Manufactured again by Henry Maudslay, there were

45 different machines in the first block making factory opened at Portsmouth dockyard in 1806. Where once 100 men had been employed, now just ten were producing 140,000 blocks a year. It could well be called the world's first mass production process.

Although there had been a good deal of mechanisation by the time vessels were being adapted for steam power, the ships themselves had changed far less dramatically. True, the Americans had produced their clippers, sleeker, faster vessels than those of the British and European shipbuilders, but a Tudor shipwright stepping aboard an early nineteenth-century sailing ship would have been impressed but not astonished by what he saw. Once the iron ship arrived, however, everything changed.

The first stage was, as with the wooden ship, the work of the designer, who laid out the lines of the vessel. The shipwright's job was once again to convert these plans into templates. It began with chalking full-size outlines on the floor of the mould loft, from which they were transferred eventually to wooden boards, where the lines were gouged out – scrieved – to form the templates from which the iron would be shaped. The next process was described in an article on Laird's Birkenhead shipyard:

> The boards are in their turn taken down to the shop below, where the ship frames are bent, on perforated plates called levelling blocks, each

The Barclay Curle yard on the Clyde in a painting by William Simpson of 1845. It shows a mixture of sailing ships and paddle steamers under construction.

> measuring 10ft by 3ft, and weighing several tons. In bending bars of angle iron for the ship's frame, a small iron mould made from the lines on the large board, taken from the mould-loft, is placed on the blocks, and its outline drawn in chalk. The mould is then removed, and after pegs have been placed at intervals along each chalk line in the holes made for the purpose, the heated bar of iron is swung on to the block by crane power, bent to the pegs, and hammered to the required shape.

Once the correct shape had been achieved, the iron went to the punching press to make the holes for the rivets, and then it was ready to be sent out to the dock where the vessel was being built.

The actual building started with laying out the berth on which the ship was to be built, with a longitudinal slope for easy launching, but an absolutely flat surface laterally. Slowly the skeleton of iron ribs was built to which the iron plates would be added. These had to be cleaned and planed and punched with holes for riveting to match those on the ribs. There were two methods of fastening the plates together, either in butt laps where they overlapped or by strapping, where the two plates abutted each other, and a third plate was riveted behind them to hold them together. Through most of the nineteenth century, all riveting was done by hand. The plates were correctly positioned. The rivets were heated on site in a brazier by the rivet boy, generally literally that, though some 'boys' never moved on to other work. Once a rivet was hot enough, it was thrown to the holder up who caught it in a scoop, lifted it out and pushed it through the hole and held the curved head in place with a heavy hammer. The two riveters on the opposite side of the plate hammered over the head with alternate blows of their hammers. It all had to be done quickly, before the rivet cooled and was no longer malleable. It was hard and brutal work and the noise of hammered metal was deafening. The noisiest place on the whole shipyard, however, was the boiler makers' shop, where walking into the building when the riveters were at work was, as one worker put it, like walking into a wall of noise.

A young worker heating rivets. He is working the bellows with his foot and is about to throw a rivet to the catcher.

By the second half of the nineteenth century, the laborious work of riveting by hand had given way to the hydraulic riveter. An account of work in the boiler makers' shop at Lairds of Birkenhead of 1874 described the process:

> In riveting boiler plates, they are suspended by an overhead crane between two dies, one stationary and the other moveable. The rivets, after being placed in the holes, are brought one by one between these dies, and the moveable die is pressed forward with a force equal to forty tons, this fixing the rivet in its place.

The early steam engines were simple, in that they had a single cylinder, from which the exhaust steam was allowed to escape into the atmosphere. The engine for the *Great Britain* was based on a design patented by Marc Brunel, for a triangular engine, in which two cylinders faced each other, inclined at an angle, with connecting rods meeting at the apex. Isambard Brunel needed a lot of power, so for the ship's engine he had four immense cylinders, each of 88in diameter, arranged in two pairs, both pairs connecting at the top of the triangle with a crankshaft. This itself was an imposing casting, with a maximum diameter of 2ft 4in. The crankshaft was connected by chains to a drum on the propeller shaft that ran between the cylinders. With this arrangement the engine speed of 18rpm was converted to a propeller speed of 53rpm. Steam pressure was only around 5psi, but

The *Queen Mary* under construction at the John Brown yard on the Clyde.

even that low pressure required a rectangular boiler 54ft long and 51ft wide, stoked from 24 furnaces. The engine produced nearly 2,000 horse power.

The low steam pressure needs some explaining, when one considers that even before work started on the *Great Britain*, Robert Stephenson had designed his locomotive *Rocket* to work at a pressure of 50psi. Even that was considered low, but the engine was built for the Rainhill trials and that had been the maximum pressure allowed. He, however, successfully tested the boiler up to 120psi. So why were shipbuilders so cautious? The answer lies in the nature of the systems. There had been boiler explosions in the early days of railway development, usually with fatal consequences for the footplate crew and damage to the locomotive. An explosion at sea, however, risked the loss of a ship, its crew and passengers. Caution was not unreasonable – and even then things did go wrong. When Brunel's third ship, the *Great Eastern*, built for passage to the Far East and Australia, set off on her maiden voyage in 1859, a mistake meant that two cocks that should have been open were closed and there was a build up of pressure in the boiler with disastrous results, as described by *The Times* correspondent who was on board:

> The forward part of the deck appeared to spring like a mine, blowing the funnel up into the air. There was a confused roar amid which came the awful crash of timber and iron mingled together in frightful uproar and then all was hidden in a rush of steam. Blinded and almost stunned by the overwhelming concussion, those on the bridge stood motionless in the white vapour till they were reminded of the necessity of seeking shelter by the shower of wreck – glass, gilt work, saloon ornaments and pieces of wood which began to fall like rain in all directions.

It was not the only alarming accident in the early years of steam. In the circumstances, caution probably seemed wise.

Improvement in engine performance was developed slowly. Experiments in the eighteenth century had shown that more efficient use of steam was possible by compounding. Instead of being allowed to escape into the air, exhaust steam which was still under pressure was passed to a second, low pressure cylinder. Randolph, Elder & Co of Glasgow built their first marine compound engine for the *Brandon* in 1854. Up to that time, the best figures for coal consumption were 4 to 4½ pounds of coal per unit of horse power per hour. On her maiden voyage *Brandon*'s consumption was down to 3¼ pounds. The significance was not lost on companies, especially those that traded to the Far East, where coaling could be problematic. Ultimately, compounding was to be taken further, with ever higher pressures, making it possible to move from simple two-cylinder compounds to triple expansion engines. As the name suggests, there were now three connected cylinders, the high pressure, intermediate and low pressure. These were the engines that would dominate the world of ocean liners through the second half of the nineteenth century.

Engine improvements also required efficient boilers. The simple boilers were little better than overgrown kettles. The inspiration for a new generation came from the railways, where Robert Stephenson's *Rocket* had a multi-tube boiler. The hot gases from the fire passed through a series of small tubes set into the main body of the boiler, heating the water. This was developed into the similar Scotch marine boiler. Towards the end of the nineteenth century, the triple expansion was being fed with steam at 125psi and the coal consumption was down to around 2lb per horsepower per hour.

The final stage was the actual launch of the vessel. Preparations for that will have begun before work got under way. Calculations would have been made to work out how high the keel would need to stand above the ground, and keel blocks then laid down. As launch day drew

The immense engine of the *Kaiser Wilhelm der Grosse* from *Cassier's Magazine* of 1898.

near, fresh calculations would have needed to be made to establish the weight and centre of gravity of the ship. Standing ways were then built up beneath the hull; then the structure down which the vessel would eventually slide was positioned and heavily greased. Wedges were put in place and, when all was ready, hammered home and the keel blocks that up to then had held the weight of the ship knocked out and all the shoring removed. Now the ship had to be held back until high tide, when the restraints were removed and, if everything had gone according to plan, the ship would slide smoothly into the water to the cheers of the crowd.

Normally, ships would be launched bow first, but there was one notable exception. Brunel's *Great Eastern*. This was by far the biggest ship of its day, and for power she had both screw propeller and paddle wheels, together with sail on six masts. She was built on the Thames but was considered to be too big for a conventional launch so she would slide into the river sideways. The day of the intended launch was 3 November 1857. The whole vessel was mounted on cradles, controlled by checking chains and winches. The great ship moved in fits and starts, until the forward cradle shot forward alarmingly, while the stern cradle appeared to have stuck. The cable on one of the winches that should have been controlling the movement broke free, sending the winch whirling, hurling off the unfortunate men who had been manning it. One died and others were injured. Everything was put on hold, and the problems took weeks to sort out. She finally made it down into the Thames on 31 January 1858.

The movement from wooden sailing ship to metal steamer revolutionised ship building. In Britain, it led to a clash between the old order and the new. For centuries, shipwrights had ruled the dockyard roost. Now they found their skills based on long years of apprenticeship and tradition challenged by a new generation of workers, who understood metal and engineering. They saw themselves as devalued and clung desperately to their privileges. But the new trades were just as keen to look to their own rights. If each group had been able to work in isolation, there would not have been a problem, but ship work cannot be compartmentalised. Shipwrights, joiners and boiler makers each demanded a monopoly on what they saw as their rightful jobs. This led to the infamous demarcation disputes that marred the British shipbuilding industry for years. Disputes could lead to bizarre decisions. For example, spars had to be added to a store room on a ship being built on the Tyne. A committee sat and solemnly pronounced that the work should go to the joiners, unless a spar was greater than 1½ inches, in which case it was the shipwright's job. One can only imagine the time wasted in wrangling over disputed

Preparing for the first attempted launch of the *Great Eastern* into the Thames. (Brunel University)

measurements. The same problems did not hinder work in other parts of the world, particularly in America, a country that had not yet built up a rigorous system based on long tradition.

One innovation did a great deal to improve hull design and make it more efficient, the use of test tanks. It was, however, an idea that had already been considered in the eighteenth century. Two great scientists, Sir Isaac Newton and Christian Huygens, had formulated the law that states that resistance to a body travelling through water increases as the square of the velocity; in other words, if you double the speed, you quadruple the resistance. This probably seemed of little practical importance at the time, as sailing vessels weren't being designed with the idea of making such a leap in speed. What is interesting, however, is that Huygens based his results on experiments made by towing objects through a tank full of water. The idea that this might be a good way of testing a hull's performance was developed in France by Pierre Bongeur in 1746 but seems not to have been taken up by shipbuilders. Ten years later in Britain, the Society of Arts put up a premium of £100 to anyone who could produce a hull design 'that would soonest pass through a given space of water driven by a given weight'. The applicants were required to supply plans and a quarter-scale model. The tests were to be very thorough, with conditions being produced to represent both calm and rough seas, and there was a test to see how close to the wind the model could sail. To ensure fairness, each model was fitted out with either a standard two masted schooner rig for a 74-gunship or a sloop rig for a frigate. Various types of test were devised, including one that involved towing the models through a tunnel full of water – a sort of maritime wind tunnel. The main tests, however, took place on two ponds. Peerless Pond in London, the capital's first outdoor swimming pool, was used for calm water tests. Frames were set up at one end of the pool, with a wheel in the centre. Lines passed over this to the models at the opposite side, and falling weights attached to the other end of the line hauled the models across, the times being recorded on stop watches. For the rough water tests, they moved to the Great Pond in Epping Forest.

The experiment was not a great success. The entrants were amateurs, who simply followed established practices. The winners were encouraged to approach the Admiralty with their designs and were simply ignored – as was the whole idea of testing hull design. The idea of using scientific ideas to influence design was revived in 1791 with the inauguration of the Society for the Improvement of Naval Architecture, but ten years later the Society was closed, having achieved nothing. Things were a little different in France, where a number of specialist papers on the science of ship building were published, but whether they actually resulted in any real differences in the ship yards is doubtful. Certainly the British paid no attention.

Things began to change in the middle of the nineteenth century, when the Admiralty established a School of Naval Architecture at Portsmouth in 1848. At least it showed that they understood that science might have a role to play in design and that they were open to new ideas. The most important new idea came, however, not from the school itself, nor even from the world of shipbuilding. William Froude had begun his career as a railway engineer, working with Isambard Brunel on the disastrous atmospheric railway that for a time linked Exeter to Plymouth. His introduction to shipping came when he also worked with Brunel on the launch of the *Great Eastern*. Perhaps this was his inspiration, for in the 1860s he moved to Paignton where he began a series of tests on model ships in water tanks. He was visited by Sir Edward Reed, the chief constructor of the Navy, who was impressed and suggested that Froude should apply to the Admiralty for a grant to set up a purpose-built testing tank. He duly sent in an application in which, among other elements, he mentioned the system then in use for testing performance, running a ship along a measured mile:

> Assuming the enquiry to result in even saving nothing better than the saving of a permanent expenditure of ten horse power in all, in the annual reduction in the national coal bill even this would pay the interest of the outlay by which the saving had been effected, were it to render unnecessary half a dozen measured mile trials in the course of each year, the same result would follow, were it to result in beneficially taking ten feet off the length of a single ironclad, the whole would probably be recouped at once.

The Admiralty agreed to pay £2,000 towards the test tank, encouraged by the loss of HMS *Captain*, sunk mainly because she was poorly designed in the first place. The new test tank was built at Torquay and soon proved its worth. The results were not lost on the commercial world. William Denny, who had a shipyard on the Clyde, established a test tank at Dumbarton in 1881, which still survives and is still in use. The test tank became an essential part in the process of designing a new ship. Just how important can be gauged by the fact that before work even started on constructing the *Queen Mary*, twenty-two 18ft models were constructed and tested to ensure that she could withstand waves as high as 50ft in the Atlantic.

The nineteenth century was to see ever bigger, faster and more efficient ships being built in the yards of Europe and America. More and more machine tools were introduced. One of the biggest changes came with a technological breakthrough. Steel production was small

One of the many tank tests used to determine the ideal hull form for the *Queen Mary*.

scale compared with wrought and cast iron. By the middle of the nineteenth century, Britain was producing 2½ million tons of iron a year but just 60,000 tons of steel. William Kelly in America and Henry Bessemer in Britain both independently came up with new, efficient methods of converting iron into steel: a form of iron with a carbon content between that of cast and wrought iron. Further improvements were made with the introduction of the open-hearth process. There was an impressive demonstration of the value of steel when the clipper ship *Clytemnestra* survived being caught in a typhoon and survived, even though the hull plates were only ⅛in thick. Other changes were to follow, which we shall look at later. But if seen from the outside, the steamers of the early nineteenth century would look to be examples of the latest technology, things could still look very different below decks. Decisions had to be taken on how the space of the hull was to be divided up, in particular the spaces that were available to the passengers. There would need to be cabins and public rooms. There was a general agreement that passengers, travelling first class in particular, would expect facilities that matched those of the grand hotels to which they were accustomed, and that implied a certain opulence. The Collins line, for example, provided public rooms full of ornate features and well-stuffed furniture, in what we now think of as the true Victorian style, never knowingly understated.

The Grand Saloon of the Collins' line ship *Atlantic*, a perfect example of nineteenth-century opulent design.

The largest ship to set sail in the early years of the twentieth century, bigger even than *Titanic,* was the *Imperator,* built for the Hamburg-Amerika line in 1913. She had scarcely had time to enter service before the start of the First World War resulted in her being laid up in Hamburg. There she remained until 1919, when she was seized by the Americans as part of war reparations and for the sinking of

Two postcards of the Cunard liner *Berengaria:* one shows a splendid modern vessel, while the first class smoking room looks as if it belongs in an old world country inn.

the *Lusitania*. She was sold to Cunard where she was renamed RMS *Berengaria* and was the flagship of the fleet until the launching of the *Queen Mary*. Technically she was a masterpiece of modern design, but the interior still harked back to nineteenth century ideals. Two postcards of the vessel illustrate the contrast: the sleek exterior and the first class smoking room, doing a fair imitation of an old coaching inn.

One might have expected liners of the early twentieth century to embrace more modern design ideas. Some did, but not all. So far, we have been looking at liners on the Atlantic routes, but as power increased, steam came to be used on other long distance routes as well. In the early nineteenth century, a British consortium began operating between London and Spain and Portugal – the two countries of the Iberian Peninsula – and became the Peninsular Steam Navigation Company. Later they extended to Italy and on to Egypt and the name was changed to the Peninsula and Orient Steam Navigation Company. More generally simply known as P & O. By the early twentieth century, the line had extended still further, running regularly between England and India and even further to Australia. The *Viceroy of India* was one of the liners built, as the name suggests, for the Indian route. Launched in 1928 from the Alexander Stephen shipyard on the Clyde, everything technological was completely up to date. She was fitted out with electric lifts between the decks and had efficient forced air ventilation where earlier ships had relied on whirling fans. She was a big ship,

A first class lounge on the P & O liner *Viceroy of India*, built in the 1920s. Although not as anachronistic as the smoking room of the *Berengaria*, it has little sense of contemporary design.

able to take 415 first class and 258 second class passengers, and also carried cargo. To look after them all was a crew of 413 that included a surgeon and his assistant. Among the new facilities was one of the first indoor swimming pools to be built on a ship. But when it came to décor, modernity ended.

Interior design was the work of the Honourable Elsie Mackay, the daughter of the P & O chairman. The music room and the dining room looked to the elegance of the eighteenth century, while the first class smoking room went further back, being roughly based on James I's palace at Bromley-by-Bow. The furniture was solidly Jacobean in style, the walls were oak panelled and the windows had leaded lights and stained-glass motifs. Dominating it all was an immense fireplace with a coat of arms above it. That most modern feature, the indoor pool, took its motif from the ancient world, inspired by the ruins of Pompeii. Other contemporary designers had very different ideas.

One year before the P & O liner was launched, a very different vessel set off on her maiden voyage on the Atlantic run. The *Île de France* was built for the Compagnie Générale Transatlantique (CGT) and was one of the most elegant liners ever built, able to carry 537 first class, 803 second and 645 third class passengers. This time the interior was not designed by the boss's daughter, but by the architect Pierre Patout. He had been involved in the famous exhibition of decorative arts held in Paris in 1925, where one of the leading furniture designers,

The smoking room of the *Ile de France*, an excellent example of Art Deco styling.

Jacques-Emile Ruhlmann, exhibited his work in a pavilion called 'L' Hôtel d'un Collectionneur', based on a house he had commissioned from Patout. The exhibition was a sensation – gone were the fussy detailing of a previous age, the heavy materials and floral patterns. In came sweeping lines and geometric shapes that took their inspiration from the work of modern artists such as the cubists Braque and Picasso. It was given a new name immediately – it was Art Deco. And that was the style that permeated the *Île de France*.

The new style was found everywhere, not just in the grand public rooms, such as the immense dining room that rose through three decks, approached down a magnificent staircase. Perhaps the greatest contrast between this ship and the P & O liner can be seen in the first class smoking room. Where the former was mock Jacobean, the French fumoir had a simple elegance and was unmistakably modern, in both style and use of materials. The style was carried through to the more luxurious cabins, where all the furniture was Art Deco and modern art decorated the walls. It set a standard that would soon be taken up by other designers, including another French architect, Roger-Henri Expert, who was in overall charge of the décor for one of the most

The dining room of the *Normandie*, surely the grandest room ever seen on any ocean liner.

beautiful ships of the age, the *Normandie*. The style again was Art Deco and Expert brought in artists such as Jean de Brunhoff, who decorated the children's dining room walls with characters from the popular *Babar the Elephant* books. The glory of the ship was the first class dining room, that rose to a height of six metres and could seat 700 guests at a time. There was no natural light, but the room had illuminated glass pillars by the master craftsman Lalique and more glass decorative columns lined the walls. Passengers entered through six metre high doors, with bronze bas-reliefs by the artist Raymond Subes. If that was all a bit too overwhelming for some passengers, they could turn instead to the café grill, which had something of the character of a Parisian brasserie and in the evening transformed into a night club.

The end of the *Normandie* was tragic for such a splendid vessel. When war broke out in 1939, she was in New York, and it was decided to convert her into a troop ship and to rename her *Lafayette*. Work was under way in February 1942, when a spark from a welding torch ignited a store of life jackets and the fire spread rapidly throughout the ship. Attempts to dowse the flames only resulted in the sheer weight of water involved causing her to begin to list to port and eventually capsized. She was salvaged, but the damage was too great. She never sailed again and was eventually scrapped.

Britain regarded it as a matter of national prestige to produce a liner to match the new generation of French vessels. Cunard reached an agreement with John Brown's shipyard on the Clyde to build the vessel of 80,774 tons at a cost of £3,992,000. She was to carry 776 first class passengers, 784 tourist and 579 third class. She too was to be modern throughout with Art Deco styling and a novelty – air conditioning throughout the first class and in the tourist class dining room. The first class had a splendid Turkish bath and swimming pool, the tourist class just the swimming pool. None of these amenities were available for the third class. All was going well, but the Wall Street crash of 1929 had caused panic in the financial markets of the world, and in December 1931 all work on the ship was halted. It was only resumed thanks to a government loan. She was finally launched in 1934, when what had previously only been known as Ship No. 534 became the *Queen Mary*. A song was written by Ina George to be sung at the event.

> There were ships of oak in the days of old
> There are ships of steel today
> And the song is the same from the men who build
> 'God speed her on her way'.
> A toast to the *Queen Mary*
> Long may she sail the sea

Here's to the name she bears
Here's to the course she steers
We want her to sail the world
With the flag of goodwill unfurled
So shout hip-hip hooray
For she's on her way
Smooth sailing to the *Queen Mary*.

She continued in service after the end of the Second World War, sailing into New York on 31 July 1947. She was retired in 1967 and bought by the city of Long Island, California, where she has been preserved as a conference centre and tourist attraction. She provides a unique opportunity to see one of the greatest of the ocean liners in all its glory. Her sister ship 'the *Queen Elizabeth*' was launched in 1938 and she made her first passenger run to New York in October 1946. She had all the elegance that one would expect from a vessel of that period. When she was retired in 1969, she was eventually bought by a Chinese shipping magnate, C.Y. Tung, who intended to use it as an international floating university. Work on conversion got under way in

A bar on the *Queen Mary*; all the art work was specially commissioned for the ship.

Hong Kong and was almost complete when fire broke out in January 1972 and the vessel was destroyed. She was the last of the great liners to be built specifically for regular passenger services.

During the war years, ship building underwent a major change. Riveting gave way to electric arc welding. There were obvious advantages in both greater efficiency and reduction in overall weight and one entire process, caulking, was eliminated. It was a process, however, that required careful planning and skilled execution. The intense heat and rapid cooling can cause distortion, and a badly welded joint would always be a source of weakness. The system was used extensively in America, and thanks to the speed at which work could be carried out, between 1942 and 1943, 27 million tons of tankers and freighters were produced and 1,238 Liberty Ships. There were initial problems, often caused by bad working practices, and some spectacular failures. The tanker *Ponagansett* was in Boston for repairs. A clip was being welded to the deck, when the whole vessel split apart, bow and stern collapsing away. Investigation showed that the problem was caused by stress patterns in the metal, and new regulations were brought in to prevent such accidents occurring again. In post-war shipyards, welding took over from riveting for the new generation of liners.

If the work of the welder was not as ear-splittingly uncomfortable as that of his predecessor the riveter, it had its own problems, as a Scottish welder explained.

> Aye it was the fumes off the welding. I don't know if you've ever seen a welder working; it's just a blue smoke comes off it. But it's alright if you're in a big place. But if you're in a wee confined compartment it just gathers and gathers and you just can't see from the one side to the other … the fumes in the confined spaces, it was terrible.

As the life of the two Queens was coming to an end, Cunard had to decide whether or not to build a replacement. They eventually decided to hedge their bets. The new vessel would be designed to meet two markets, the declining transatlantic passenger service and the rapidly growing demand for luxury cruising. Instead of the three classes of the old liners, there would just be two – no one would want to go third class on a cruise. It was soon apparent that it was cruising that would keep her afloat. Fashions had changed again. The second *Queen Elizabeth*, generally known as the *QE2*, was launched in 1972 and reflected the design ideas and materials of the time. Aluminium and stainless steel were on display, Perspex and laminates were widely used. The overall effects were of simplicity rather than extravagance. The Queen's Room for first class passengers typified the style, with muted colours of white and tan, and a lowered ceiling with indirect lighting. Columns were flared as they reached the ceiling. By contrast the Theatre Bar was a symphony in red – red furniture, red curtains and even a red piano. The artwork spread through the public rooms was uncompromisingly modern, including several abstracts. It was stylish but perhaps lacked the grandeur of earlier liners.

The finest of all the post-war ships in style and performance is arguably the SS *United States*, that was built in 1950-51. With memories of the war still very much alive and tensions existing between east and west, she was specifically designed so that she could, if necessary, be easily converted into a troop ship, a function that happily she never needed to perform. The designer Dorothy Marckwald, who had earlier worked on designing the pre-war SS *America*, was to follow the same basic, modern ideas, but the dual nature of the ship dictated the choice of materials to a large extent. A troopship needs to be fast and, given that it is likely to be under attack, fireproof. There was no question of using traditional wood panelling. Instead, aluminium and fibreglass were used, while upholstery was in the artificial fibre, Dynel. The banning of wood even went as far initially as a proposal that the grand piano in the ballroom had to be made of aluminium as well. Fortunately for music lovers, when it was shown that even when dowsed in petrol that was set alight, mahogany refused to burn, it was agreed that a more conventional piano could be used after all. The 26 public rooms, 674 state rooms and 20 luxury suites, were all elegantly simple. Colour schemes were light and subtle and decorative artwork

The 'Expression of Freedom' sculpture on the *SS America.*

emphasised American themes. One of the most striking was a monumental sculpture, 'Expressions of Freedom' by Gwen Lux. The good news is that, after retirement, there are now plans to restore the ship as a permanent cultural centre, and while many of the original internal fittings have gone, any development will have a museum giving the ship's history.

Looking at the construction of ocean liners is to get a view of the changing pace of industrial design and development. The interior is a mirror of changes in architecture and the decorative arts. There is scarcely a style that did not find a place somewhere in the liners of the nineteenth and twentieth centuries. Designers borrowed from everywhere, from ancient Rome to Moorish Spain and from the Baroque of Louis XIV to the sleek modernism of Twenties Paris. A few of these have been described here, but they have been chosen to give as good a view as possible of the great diversity that Atlantic passengers once enjoyed.

Chapter Five

Fitting Out

There were several ways to attract travellers to take the new steamships when they needed to cross the Atlantic. The most obvious one was that they were quicker. It was not simply a matter of saving time, but regardless of what brochures might promise, an Atlantic voyage was not necessarily an enjoyable experience. In the 1950s, I had been called up for National Service and was lucky enough to be appointed for aircrew training in Canada. We left Liverpool for Halifax, Nova Scotia in a comparatively small cargo-passenger vessel of the Furness-Warren line – small but large compared with the early nineteenth-century steamers. It was February and we had hardly left the mouth of the Mersey and were enjoying the first meal on board when one of the passengers turned pale and left the table, not to reappear until we were in sight of the Canadian shore. Later, we hit rough weather. For those with sound stomachs, it was quite entertaining being in the bar trying to catch your glass as it slid from one side of the table to the other, but for many it was pure misery. The majority on that ship would have been heartily wishing that the whole voyage could be ended as quickly as possible. Speed then was a good selling point. But it was also the job of the designer of the state rooms and saloons to make everything as comfortable as possible. One designer was asked why he did not use more nautical themes in the main rooms, to which he replied that the last thing many passengers needed was to be reminded that they were at sea. As a result, designs were based on familiar rooms on the land – rooms that did not pitch up and down or roll from side to side but remained firmly rooted to the ground.

There was another aspect to transatlantic travel that affected design decisions; as a passenger what you got depended on what you could pay. The first class wanted to see and be seen in appropriately grand surroundings. Needless to say, those travelling in lower classes had lower expectations, and accepted more spartan surroundings. So, when it comes to design, the emphasis is very much on the grandeur rather than the mundane. This was where the designers made their mark, though it took some time before the ideals of comfort and splendour were actually achieved.

Once Brunel had proved that steam could conquer the Atlantic, others were bound to follow. In March 1839, the engineer Robert

The first class dining room on the *Balmoral Castle.*

Napier noted a meeting that was to mark the start of one of the great ventures of maritime history:

> Mr Cunard from Halifax [Nova Scotia] came here yesterday afternoon, and I have this day contracted with him for three vessels to carry the mails from this country to Halifax. He appears from the little I have seen of him to be a straightforward business man, and from all I can judge he has made a very favourable contract.

The vessels were paddle steamers, each to be not less than 200ft long and 32ft between the paddles and to have 'cabins furnished in a neat and comfortable manner for the accommodation of from sixty to seventy passengers'.

Another magnificent dining room, this time the *Lusitania.*

One of the three ships was the RMS *Britannia* and among its early passengers was Charles Dickens, who described the experience in his book *American Notes.* He had high expectations of what he would find when he got aboard, lured by the 'highly varnished lithographic plan hanging up in the agent's counting house in the city of London'. State rooms were 'chaste and pretty, not to say gorgeous little bowers' while the saloon was a 'chamber of almost interminable perspective, furnished, as Mr. Robinson would say, in a style of more than Eastern splendour, and filled (but not inconveniently so) with groups of ladies and gentlemen, in the very highest state of enjoyment and vivacity.' The reality was very different. Before going down to the state room reserved for himself and his wife, he got to see the saloon:

> Before descending into the bowels of the ship, we had passed from the deck into a long, narrow apartment, not unlike a gigantic hearse with windows in the sides; having at its upper end a melancholy stove, at which three or four chilly stewards were warming their hands; while on either side, extending down its whole dreary length, was a long, long table, over each of which a rack, fixed to the low roof, and stuck full of drinking-glasses and cruet stands, hinted dismally at rolling seas and heavy weather.

The encounter with the space which was reserved for them down below came as a shock:

> I shall never forget the one-fourth serious and three-fourths comical astonishment with which on the morning of the third of January eighteen-hundred-and-forty-two, I opened the door of, and put my head into, a 'state-room' … That this state-room had been specially engaged for 'Charles Dickens, Esquire and Lady', was rendered sufficiently clear even to my scared intellect by a very small manuscript announcing the fact, which was pinned on a very flat quilt, covering a very thin mattress on a most inaccessible shelf.

It was an 'utterly impracticable, thoroughly hopeless, and profoundly preposterous box' and as he sat on 'a kind of horsehair slab, or perch' he hoped that it was just a 'pleasant fiction and cheerful jest of the captain's, invented, and put into practice for the better relish of the real state-room'. Sadly for Mr. and Mrs. Dickens, it was not a joke, but would be their accommodation for the voyage. This is not the image that comes to mind when the name 'Cunard' is mentioned, but this was only the beginning of a line that, in time, would produce some of the most gorgeously equipped vessels to cross the Atlantic.

When Collins began competing with Cunard, his early ships, the *Atlantic* and the *Pacific*, were fitted out in a sumptuous style that might have satisfied the fastidious Charles Dickens. There were two saloons, each 67ft long and 20ft wide, with expensive panelling in rosewood, satinwood and white holly. One was the dining room, but it was the other, the drawing room, that attracted the admiration of visitors, described by one as 'the most beautiful and gorgeous apartment of the kind that was ever seen', with comfortable upholstered chairs and sofas, marble topped tables and thick pile carpeting. Mirrors hung round the walls and the windows had emblems representing American freedom. Everywhere there were symbols of Americana, with the coats of arms of the states and an oil painting which, in case

The opulent saloon on the *Great Eastern.*

any of the passengers had any doubts on the subject, emphasised the superiority of that country over those of the old European monarchies. American Liberty was stomping on a feudal Prince. There were many features that were quite new for ships. Where Dickens had to make do with a stove in the saloon, the rooms here were steam heated. There were bathrooms, though bathing in a rough sea must have been an interesting experience, and a barber's shop. One set of features that would certainly not have been seen or welcomed on a European ship were the spittoons. A feature that would be common on later vessels was a smoking room in an outhouse on the decks, but it lacked any of the opulence of the other public rooms.

America expected great things of the Collins vessels and they were not disappointed. Over the few years, speeds would increase and the interiors of ocean liners would become ever more opulent. Brunel's *Great Eastern* was the largest ship of its day – and was to remain the largest for many years. She was also fitted out with a splendour that could match that of the finest hotel. The public rooms were magnificent. The main saloon had ornate stuffed sofas, covered in rich velvet, gilded pillars, patterned carpets and windows, with ornate drapes. Where a funnel had to pass through the room, it was disguised behind panelling. More light was supplied by immense skylights and the walls were festooned with mirrors. The main dining room was equally grand, with a high arched ceiling of red and blue and it even boasted a piano at one end on a platform. The smoke room was very far from being the rough shed on deck of the Collins ships. Here it was as fine as the other public rooms, with fourteen windows and a ceiling of red and gold.

It was not just the space below decks that set the *Great Eastern* apart from other vessels of the time. She had a promenade deck stretching both sides of the vessel, where passengers could saunter in good weather. Jules Verne travelled on the ship and used it as the setting for his novel *Une Ville Flottante* of 1874 – *A Floating City*. In it, he described the scene on board in the afternoon on a day when clearly the sea was calm:

> Lunch over, the decks were again filled, people bowed and spoke to each other in passing as formally as if they were walking in Hyde Park; children played and ran about, throwing their balls and bowling hoops as they might have done on the gravel walks of the Tuileries; the greater part of the men walked up and down smoking; the ladies, seated on folding-chairs, worked, read, or talked together, whilst the governesses and nurses looked after the children.

Dickens would have been delighted to find that if he had booked a state-room on this vessel, because he would actually have got

A deck scene on the *Great Eastern* by Robert Charles Dudley.

something that met the description. Apart from the beds, each had a rocking chair and a sofa, the latter concealing a bath tub. There was a separate wash basin with hot and cold water on tap. With a richly carpeted floor, it was a room that offered genuine comfort.

Because the great ship was not intended for the Atlantic trade, but had been built specifically for journeys to the Far East and Australia, every effort had to be made to ensure the passengers had a wide range of amenities. On such a long voyage, the comforts she offered would have been much appreciated, but she was never to be used on that route. Her maiden voyage in June 1860 was with a crew of 418 and just 35 paying passengers. The largest number she ever carried was on a voyage from Britain to Quebec, during the American Civil War. The British government decided to send 2,000 troops to Canada together with 122 horses and more ordinary passengers. The following year, however, she was caught in an immense storm, with hurricane force winds, when just two days out of Liverpool. Her paddle wheels were destroyed and her sails shredded, leaving just the single propeller to keep her moving, but her rudder was bent and when that hit the prop, all power was lost. An American engineer on board managed to rig up a temporary propeller, and eventually she was towed back to port.

She was never a success as an ocean liner, in spite of her opulence, but she was to have a vital role in improving communication around the world. In 1862, she was fitted out for laying a telegraph cable across the Atlantic. She set sail from Sheerness in May 1866 and by July of that year the work was complete. Messages that had once taken days or even weeks to be transmitted between North America and Europe could now be sent in minutes. The vessel was not the success her builders had hoped for as a passenger ship, but she had proved her value at the end. Her cable laying days over, there was an attempt to return her to passenger duty but was never a success. The final ignominy came in 1885 when she was auctioned off and bought for just £26,200 and was for a time used as a show boat, staging concerts, before the final insult, when she was sailed up and down the Mersey as a giant advertising hoarding for Lewis's department store. She was sent to the breaker's yard in 1888.

By the latter part of the nineteenth century, designers were employed specifically to produce interiors that would match the taste of the period. Cunard ordered two identical fine vessels, the *Campania* and the *Lucania*, at the time the *Great Eastern* was being scrapped. These were thoroughly modern ships. Sails were no longer needed, as each was fitted with two immense triple expansion steam engines driving twin propeller shafts. The vessels themselves had graceful lines and a pair of raked funnels, looking quite sleek and modern. Inside, things looked very different, with styling that looked more to the past than the future. The interior was a monument to Victorian opulence. Heavy panelling covered the walls of state-rooms and public rooms. The finest of the former offered both a bedroom and a parlour, with a brass bedstead and a hanging wardrobe in the one and the latter, in the words of the company's own brochure, was 'fitted up with tables and chairs after the style of a lady's boudoir.' In the public rooms, there seemed to be no need to have any sort of consistent style. The first-class entrance hall for example, was French Renaissance, while the men's smoking room was Elizabethan English. The latter had the solid respectability of a gentleman's club, while the ladies' saloon was all pretty flowered textiles and carpets. The grandest room of all was the dining hall, an immense room 10ft high, 63ft wide and 98ft long. This time the style was described as 'modified Italian', which translated into a rich mixture of mahogany panelled walls, inlaid with ivory and separated by decorated pilasters. The central part rose through three decks to a skylight and the white and gold ceiling was supported on ornate classical pillars.

The great rooms are what we all remember when we think of the mighty liners of the past, but fitting out involved far more than that. Everything had to be provided that might be needed on the voyage. The larger liners would have a surgeon and medical room on board, for example. There were trades that one does not often think of as being involved in ship building, but plumbers had to be employed to install wash basins, baths and toilets, just as they would have been for a new hotel. Carpenters provided built-in furniture and armies of decorators were employed. Some work went to outside contractors. On the *Queen Mary*, for example, the contract for the first class swimming pool and the Turkish baths, for £10,498 and £4,050, went to Trollope & Sons, while the far less ornate second class pool was built by Wylie & Lockhead for a more modest £7,174. But there were also a whole variety of smaller items that had to be provided – there was little point in having the grandest first class dining room if plates and cutlery were dull and plain, so most companies had these items specially made, often incorporating the Company name and crest. These were vast orders – the *Queen Mary* after all had a population bigger than most villages, with a grand total of over 2,000 passengers and 1,100 crew. Fitting out a liner was an immense operation.

Chapter Six

The Crew

The undisputed head of all things on board ship was the captain. In the days of sail, there was a marked difference between the captains of American ships and British. The former were usually either owners or part owners of the vessels in which they sailed, which was rarely the case with the British. This was, no doubt, one reason why the Americans were generally considered to be the best in the business – even by the British themselves. There is evidence of the high esteem in which the Americans were held in evidence given to a select committee of the House of Commons set up in 1841 to look at the conditions in which Irish emigrants were carried across the Atlantic. One factor was that whereas English insurance companies were entirely concerned about the physical state of the ship, American insurers were equally concerned about the man in charge of it. 'They know all about him, his history, where he comes from, and what his habits are; and they will not insure a vessel, except at very high premium, unless they are satisfied the captain is a decent man.' This was sensible, as ultimately the safety of a vessel or its wreck could well depend on the skill of the man in charge in dealing with severe conditions at sea.

Terry Coleman in his book *Passage to America* (1972) gives an example of how a captain might show off his skill. If the wind was in the right direction, he would leave his berth in the harbour by setting full sail and heading out stern first into the main channel, before spinning the ship round and heading out to sea. There do not seem to be any authenticated accounts of this happening but sailing off a quay was certainly possible as I discovered one morning on Orkney. Two square rigged vessels, the barque *Marquesa* and the brigantine *Cuidad de Inca*, both sailed off the quay at Kirkwall. It was a memorable sight; tragically the *Marquesa* was lost at sea, a reminder that even in this age with all the modern navigational and other aid available, sailing ships are vulnerable. The nineteenth century packets were even more at risk.

The captain was, in a sense, the overall strategist for the voyage, setting the course, deciding when to tack, when to set more sail and when to take it in. The orders were relayed to the crew via the mates. The helmsman would be given the compass course to steer. The officers would be checking position whenever possible – and it was impossible when the sky was overcast. If those conditions continued for days on end – and on the Atlantic that was always a possibility – then the ship could be drifting off the true course. This was not necessarily a problem when the vessel had a favourable wind and the captain could be reasonably confident that the course he had set from the chart would be followed with a good degree of accuracy. The same did not apply in other circumstances, when the ship was regularly tacking to make way, or when a heavy swell was running. It was in such circumstances that experience and good judgement were essential.

Captain Janis Drezihs, looking suitably imposing, in a studio portrait; he commanded several ocean going vessels, beginning his career on sailing ships.

It was not the job of the captain only to ensure the safety of the ship under his command, he was also responsible for the welfare of the crew and passengers. Some captains were benevolent but by no means all, and the worst seemed to be those on the ships specialising in the immigrant trade. Nathaniel Hawthorne is best known as a novelist but in 1853 he was appointed American Consul in Liverpool, where he took a keen interest in the Atlantic trade and the men who worked the ships and wrote several reports on conditions aboard. His early views of American captains were generally favourable when he had first met them, but the brutalities of the immigrant trade had made things seem very different. In 1857, when he left Liverpool, he wrote:

> I deeply regret to say, that the character of American shipmasters has already descended, many degrees, from the high standard which it has held in years past; an effect partly due, as I have hinted, to the constantly narrowing field of selection, and likewise, in a great degree, to the terrible life which a shipmaster is now forced to lead. Respectable men are anxious to quit a service which links them with such comrades, loads them with such responsibilities, and necessitates such modes of meeting it.

An example of a captain's treatment of the immigrants on board was quoted earlier but the treatment of the crew was often just as brutal. The owners demanded that their ships were driven hard; the shorter the voyage time, the greater their profits. Experienced crews could manage this with efficiency but it seems that not all crews were capable. A man could claim to be an accomplished seaman, but there would be no documents to back up the claim. It seems, from evidence given to the migration commission of 1854, that when men were signing on for vessels bound for Australia, if there was any doubt about the man, then he would be sent aloft – it would soon be clear whether he was experienced or not. A man who had sailed before and been aloft in rough weather, would have no problem clambering up the rigging and moving out along the yardarm of a ship in harbour. This, it seems, was an ability that was never lost. A neighbour of mine told me about taking his father, then in his eighties, to see the tall ships on display at Gloucester Docks. He had sailed in square riggers in his youth, and when he went on board one of the vessels, he asked if he could go aloft for old times' sake. Given permission, he climbed to the top yard and then shinned up to the top of the mast, took his hat from his head, popped it over the top, waved, put his hat back on and came down as nimbly as he had gone up. No captain would have hesitated to sign up a man who did that.

The immigrant ships attracted the worst masters and the poorest crew. Life on board could be brutal. In good weather with a following wind, everything could go smoothly. The crew would finish their watch and hand over to the next, and would then be able to have something to eat and a rest before they were back on deck again. But in bad weather they might have scarcely settled down in the fo'c's'le before the call for all hands on deck rang out, as the ship was about to tack. Some might be sent aloft, while others took to the lines, moving the spars. This was hard work and it was essential to work as a team. The best way of achieving that was to have a song to provide the rhythm. These songs are what we now know as sea shanties. One of the best known is 'Haul the bowline' with its refrain, repeated after each line – 'Haul on the bowline; the bowline haul'. On the word 'haul', the men would pull. Many versions exist, some of which include the line 'the captain was a-growling' but it seems that this was not always the case – and might earn the wrath of the ship's officers if they saw it as an insult. One unhappy sailor who added the line was knocked over the head with a bucket by one of the mates, and when he was down, battered with another bucket by a second mate.

It could take as long as half an hour to complete a tack, and there was no question of waiting for fine weather; if it was a howling gale and pouring rain, when the order came then it had to be obeyed. And it was not just when tacking, there were constant adjustments to be made to the sails to make the best of whatever wind was available or to limit the damage if it became too strong. In winter, the rigging could have a coating of ice, making it doubly dangerous. Herman Melville wrote a novel, *Redburn: His First Voyage* (1849), in which he describes a young man 'Wellingborough Redburn's' first trip as an ordinary seaman between America and England, based very much on his own experiences. So, although it is a work of fiction, one can be reasonably sure that what he writes is accurate. He described many of the activities on board. In one particularly passage he describes how, while still inexperienced, he had to go out at the dead of night to loosen the main sky-sail. This was the highest sail on the ship, the fifth up from the deck. He clambered up the rigging, until, looking down, 'I could but just perceive the ship below me, like a long narrow plank in the water; and it did not seem to belong at all to the yard, over which I was hanging.' He loosened the gaskets, the lines holding the sail to the yard, and shouted 'Hoist away':

> And hoist they did, and me too along with the yard and sail; for I had no time to get off, they were so unexpectedly quick about it. It seemed like magic, there I was, going up higher and higher, the yard rising under me, and not a soul in sight. Without knowing it at the time, I was in a good deal of danger, but it was so dark that I could not see well enough to be afraid – at least on that account; though I felt frightened enough in a promiscuous way. I only held on hard and made good the saying of old sailors, that the last person to

Under sail in heavy seas: the men out on the yardarm are probably taking in the sail.

> fall overboard from the rigging is a landsman, because he grips the ropes so fiercely; whereas old tars are less careful and sometimes pay the penalty.

At times, there were other hazards to contend with. The immigrant ships often struggled to find return cargoes once they had landed all the passengers, so they had to return with ballast in the hold. In bad weather, that could shift and if it moved too far, the vessel would start to heel over. Then there was nothing for it but for some of the crew to descend to the dark, wet bowels of the ship and move it – possibly sand to be shovelled or rocks to be carried. The crew earned what little pay they got.

There were always tasks to perform, as Melville described:

> At sea, the sailors are continually engaged in 'parcelling', 'serving' and in a thousand ways ornamenting and repairing the numberless shrouds and stays, mending sails, or turning one side of the deck into a rope-walk, where they manufacture a clumsy sort of twine, called spun yarn. This is spun with a winch: and many an hour the Lancashire boy had to play the part of an engine, and contribute the motive power. For material, they use odds and ends of old rigging called 'junk', the yarns of which are twisted into new combinations.

This remarkable photo shows a desperate sailor, clinging onto the yard after the rope on which his feet had rested snapped. No details were given of the ship on which the accident occurred.

Even in the sailing days, when the immigration trade was at its height, there were well managed packets that carried a large complement of paying passengers in the first class accommodation. They required extra crew to look after the cabins and their occupants. The cooks had to be able to provide good, well-cooked food and stewards had to be available to serve it – and serve whatever drinks or snacks passengers might want right through the day. A less savoury duty would be seen in the mornings, when the stewards lined up on the lee side of the vessel to empty the chamber pots from the cabins. The same courtesies were not, needless to say, extended to the emigrants in steerage. In other words, the ship had to provide all the amenities of a decent hotel and have the staff to ensure it was well run.

Steamships required a new form of crew, previously unknown on any ship – the stokers and engineers who worked down in the bowels of the ship, with the engines and boilers. For them, the hardest and dirtiest work began even before the ship set sail – bunkering, taking on board the coal. For the big ships of the early twentieth century, this was a vast amount. The bunkers of *Titanic,* for example, were designed to hold 6,611 tons. Before coaling could begin, everywhere that was open to passengers had to be sealed off – ventilators closed, portholes shut and every door that connected with the open air or the boiler room shut. After that, the ship would be moved away from the quay to allow the coal barges to come alongside. In a description of coaling at an American port, the barges carried anything from 500-1,000 tons each. The coal was loaded through openings, coal ports, a few feet above the waterline. These had been tightly sealed during the previous voyage, and now it was the task of the ship's carpenter to lower himself overboard to open the heavy iron doors. In a typical operation from around the end of the nineteenth century, derricks would be set up above each coal port, from which buckets, each holding around a quarter of a ton, could be hauled up from the barge to men on suspended small platforms beside each port. They emptied the buckets into chutes, down which the coal shot down into the boiler room bunkers.

Inside the ship, were the trimmers. Their job was to move the coal from the bottom of the chutes to where it was needed, using shovels and barrows. This was an operation that had to be executed very carefully under the direction of the chief engineer. It was essential that the coal was loaded and spread equally to prevent the vessel listing to one side. The trimmers generally worked in gangs of six, shifting ten to fifteen tons an hour, in clouds of coal dust, with no more than a wet rag round the mouth for protection. It was cruelly hard work and coaling could take up to two days to complete. Once everything

Bunkering a liner at Hoboken on the Hudson River, opposite New York. Coal is being loaded into the ship from the barge.

had been completed and the dust had settled, the rest of the ship could be opened back up again. The carpenter was back over the side, ensuring that all the coal ports were firmly closed, as they were near enough to the waterline to be submerged in a heavy swell. By an oversight one of the coal ports on the *Mauretania* was not properly sealed on a voyage during the First World War. Water poured in faster than the pumps could remove it. As the ship began to list, there were frantic efforts to reseal the opening. They were successful, the water stopped flowing and what was left was removed and the ship sailed safely on.

Once the coaling was complete, there was little time to rest for the trimmers, as the ship prepared for sea, and they now began working with the stokers. Once again, the quantities of coal that had to be shifted were prodigious: the *Titanic* burned as much as 800 tons a day. The ship had 29 boilers, in which the water was heated by 159 furnaces to provide steam at a pressure of 215psi. It is a measure of just how far the ocean liner had come since Brunel's *Great Western* crossed the Atlantic. The engine there worked at a pressure of just 5psi. Keeping all this going on the *Titanic* required a massive effort by the stokers and trimmers – and it was a large work force. There were 13 leading firemen or stokers, overseeing a crew of 163 stokers and 75 trimmers, collectively known, for obvious reasons, as the 'black gang'.

Once under way, the trimmers had two main jobs: the first was wheeling barrow loads of coal from the bunkers to the stokers and carrying away the ash; the other was keeping the coal on the bunkers trimmed, just as they did when loading. On the big liners, the stokers worked four-hour shifts, which was as much as a man could manage of continuous hard labour in a temperature that could soar to 50°C. They would each be responsible for three or four furnaces, and their task was to ensure that the steam pressure was kept at the required level by keeping a bright, evenly burning fire. This involved a rota of raking out the clinker left behind by the burned fuel, levelling the fire and throwing in the appropriate number of shovelfuls of coal with precision. The speed to which they worked was set by the engineer in charge and depended on the conditions the ship was facing and the speed required. Everything had to be done in the correct order and the efforts of all the stokers co-ordinated. Throughout the nineteenth century, this required some sort of control, such as the ringing of a bell to indicate when the stoker should carry out each sequence of operations. In 1903, the process was mechanised by an electric device, named after its inventor, the Kilroy Stoking Regulator, several of which would be installed. Visual indicators showed which furnace was to be fired next, and gongs gave the time period in which the furnaces had to be fired, before the next cycle began. For the stokers, this meant manhandling the heavy gear needed for removing clinker and

The 'black gang' – coal trimmers and stokers at work.

levelling and throwing in the coal, generally at 10lb a time. The shorter the time set on the regulator, the faster they had to work.

It was brutally hard work, and there was little in the way of comfort in the breaks between shifts. The accommodation for the men who worked in the boiler room was generally the poorest in the ship, set right up in the bows, so that if the ship was pitching, they got the worst of the movement. They slept in tiers of bunks in a dormitory, and in spite of the filthy coal dust with which they were covered, the facilities, even on modern vessels such as the Olympic Class, they had neither baths nor showers. To keep them fed, the ship employed six mess stewards, who cooked and served the meals.

Not surprisingly the black gang had a reputation as hard men – drinking heavily when they had the chance which, in theory at least, was only when they went ashore. Fighting was not uncommon, and it was said that it was unwise for an engineering officer to enter the boiler room alone, as there was always the possibility that he would be coming out feet first. Pay was not high, and they were signed on for the voyage only. So once the ship docked, they received their pay and could either wait until the ship was ready for the return voyage or try to get a berth on another ship. The trimmers were more fortunate, in that they still had work to do during the turn round period and got paid.

Given the low status in which stokers were held and the nature of the men and the job, it is hardly surprising that trouble broke out on board from time to time but was rarely more than an outburst of anger and was quickly settled. The exception came on the Cunard liner SS *Ultonia,* built in 1898. The stokers refused to work, and what would have been a strike on land was a mutiny at sea. The men were confined to their quarters and ordinary sailors sent down to the darkness to take over the stoking. They did their best, but their efforts could not keep the furnaces going flat out and the ship crawled along at a miserable 2 knots.

The world of the stokers and trimmers only came to an end when oil began to be used instead of coal for the furnaces and when diesel engines began to replace the steam engines that had dominated the liner trade for a century. But as long as they were needed, they were without doubt the hardest worked of any on board any ocean liner.

Although engines had got ever bigger and more powerful, looking after them required only a small fraction of the crew necessary to keep them in steam. Until the end of the nineteenth century, all the engines had been reciprocating, with pistons moving up and down in the cylinders. As steam pressure had increased, so it had been necessary to make the most of it, and it seemed that the ultimate efficiency had been reached with the triple expansion engine. Then, in 1897 the Norddeutscher Lloyd Company of Bremen launched the *Kaiser Wilhelm der Grosse* with a quadruple expansion engine, in which the steam entered the high-pressure cylinder, then when exhausting passed through two intermediate cylinders to the huge low-pressure cylinder, before finally entering the condenser to be turned back to liquid water. This was the biggest engine ever seen on an Atlantic liner and gave the ship an impressive speed of 22 knots but was soon to be followed by the *Kaiser Wilhelm II* of 1903, with two of these massive engines driving twin propellers.

The chief engineer was very much the man in command of the engine room and there seems to have been a long tradition in British vessels of the job going to Scotsmen. The chief was responsible for keeping the engines in perfect working condition and ensuring that orders that came down from the bridge were carried out promptly and efficiently. Steam pressure had to be maintained at the appropriate level throughout the voyage, and a small team of greasers was responsible for keeping the engines regularly oiled and running smoothly.

Communication between the captain on the bridge and the chief engineer was always crucial, and complicated: the captain could see where the ship was going, but had no idea of what was happening in the engine room; while the engineer knew how his engines were performing but being below the waterline had no idea of where the ship

C.W. McKimm, the Chief Engineer of the *Olympia.*

was heading. Until the late nineteenth century, the best available means of communication was the speaking tube. Then in 1870, a Liverpool marine instrument manufacturer 'William Chadburn' introduced the engine order telegraph, often known as the ship's telegraph. The telegraph on the bridge has a dial and a handle that moves a pointer to indicate the speed from full ahead at one end of the dial to full astern at the other, with stop and finished with engines in the middle. This sends an electrical signal to an identical device in the engine room and a bell sounds to announce the order has been given. The engineer then moves the handle to adjust the needle to the corresponding position to acknowledge the message has been received and is being acted on.

Few passengers saw anything of this world below decks. A few interested parties might be allowed down into the engine room to see the imposing machinery, but none was permitted into the boiler room. Partly, this was so as not to interfere with the smooth running of the operation, but the fine ladies and gentlemen of the first class would probably have ended up suing the company when they emerged black of face and with clothes ruined.

Things were very different above decks. The captain of a steam liner held a master's ticket and was indeed master in his floating kingdom, ultimately responsible for everything that had happened to the ship. His duties began before the ship set sail, when he would make a thorough inspection to ensure everything was ready and in good order. All the necessary paperwork needed to be checked, and he had to ensure that the ship's officers had carried out their shore duties, such as ensuring that all cargo was safely and efficiently stowed. He would make sure that all necessary stores were on board and would check the passenger list to see if there was anyone whose status required a little extra attention. If there was, he would make a note of that and, next day when they came on board he would be at the top of the gangway and personally escort them to their quarters. In many cases as the vessel left harbour, the captain would hand temporary control of the bridge to the pilot until the ship was clear of any dangers, at which point the pilot boat would come alongside to take off the pilot and leave the

A captain standing beside the ship's telegraph.

captain back in full control until the end of the voyage, when the next pilot might be needed to see them safely into harbour.

The captain set the course for the day, but this would be subject to regular checks and, if necessary, amendments. His decisions were based on the results of the navigation officer of the day's readings. As on the sailing ships, throughout the early years of Atlantic steam, navigation 'once land had been left behind' depended on readings taken of various astral bodies. As explained earlier, latitude could be found by taking a sun shot with a sextant at noon. But other readings were also possible. *The Nautical Almanac* was first published in 1767, but the version used by steamships would have been developed from the revised version of 1834. It consisted of a whole series of figures relating to the moon, planets and certain stars, showing the Greenwich Mean Time when they would be overhead. By taking the angle at which they appear when measured by the sextant, a line can be drawn across the chart for that time. By taking two, or better still three, measurements of different bodies, you can get three lines, which, when making a slight allowance for the ship's movement, should all intersect at a precise point, which is the ship's actual position. If that point is not on the line corresponding to the course set by the captain, it means that current and wind have moved the ship away, and a new course can be set.

The job of the navigation officers was crucial for a successful voyage, and it was no easy task learning the skills needed. Using a sextant on land is comparatively straightforward, but it is a different matter trying to keep a level with the horizon and finding the appropriate star when the vessel is heaving around beneath your feet. There was a further problem on the early liners, where the bridge really was just that – an open structure with little or no protection from the elements. Navigational errors could be costly. J. Murray Lindsay in his book *Sailor in Steam* (1966) described an incident that occurred when he was mate on the *Takliwa* in 1924. This was a brand new ship, launched that year in Glasgow and about to embark on her maiden voyage from India to Japan. They were just leaving for the return voyage out of Yokohama, when she ran aground:

> It took only a few minutes' study of the chart, and some bearings from the misty land, to pinpoint where, and how, we had gone ashore. In the Uragakaiko Strait, the bottleneck between the inland sea of Tokyo Wan and the Pacific, lie two small islands dividing the entrance into three channels. Confused by the fog, we had entered the wrong channel, thereby bringing to an untimely end a long sea career for a Captain we had all grown to admire as a sailor. There is no second chance for men of the sea.

On the return to India, the Captain was, indeed, to lose his position. It could all have been a lot worse. The ship was saved. Fortunately, although the bows were firmly stuck, the stern was in deep water, and after several attempts, with the mighty quadruple engine going Full Astern, she gradually pulled out and could make her way without any further problems. But this was the sort of incident any captain dreaded, since, as Lindsay wrote, there was no second chance.

One of the more common causes of accidents at sea was collision. In theory, the ship's officers were responsible for keeping a watch night and day. The most dangerous time was at night, when the last passengers had retired, the public rooms were empty and lights switched off. All ships carried navigation lights, green on the starboard and red on the port sides. A good look out would be well able to see another vessel approaching. But those night hours are also the ones where drowsiness can take over. If things go wrong, it is the job of the duty officer to take appropriate action. The rules of the sea are clearly laid out and each would-be officer is tested on his knowledge of them, before he is given his commission. Lindsay recorded an alarming incident, when he took on his first job as mate, having passed the tests and learned the rules.

It was night and the captain was down below, fast asleep, leaving him in charge. The lookout spotted another ship advancing towards them. He remembered the rules he had learned, including Article 19. 'When two steam vessels are crossing, so as to avoid risk of collision, the vessel which has the other on her starboard side shall keep out of the way of the other.' In this case, it meant that Lindsay should hold his course, and the other vessel's officer should change his. And that is exactly what Lindsay did with almost disastrous results.

> The lookout raced down from the fo'c'sle head as a large cargo ship with all lights blazing sheered across our bows to clear us by a bare few yards – and it really was a matter of yards. I saw no one on the bridge and she did not answer my signals by morse lamp. She was so close I could see the pictures on the walls of her cabins, and we rolled in each other's wake.

Unfortunately, Lindsay had, in his inexperience, forgotten another rule – Article 27. That one stated that in exceptional circumstances, all other rules should be ignored and the necessary action taken 'to avoid immediate danger'.

The captain was never officially off duty. If anything occurred that needed his attention he would be summoned, day or night. He might be entertaining important passengers at his dining table when a discreet note would be passed, and he would make his excuses and leave, or it might be the middle of the night when he was asleep, but he would need to be woken. He was not only responsible for the ship, he was expected, on the great liners, to be host to the passengers. In her book *The Captain's Table* (2005), Sarah Edington quotes Commander Gordon Steele:

> Every Captain of a liner is supposed to be a 'character' otherwise he would not attract passengers to his ship. It does not matter what form his peculiarity takes. I have known rude Captains, polite Captains, Captains who are fond of children, Captains who are fond of ladies and Captains who are fond of neither. I have sailed with liberal drinking Captains and teetotal Captains.

The captain would find time to chat, even late at night and in bad weather, as Dickens recorded:

> Eleven o'clock or thereabouts, when the captain comes down again, in a sou'-wester hat tied under his chin, and a pilot-coat; making the ground wet where he stands. By this time the card-playing is over, and the bottles and glasses are again upon the table; and after an hour's pleasant conversation about the ship, the passengers, things in general, the captain (who never goes to bed and is never out of humour) turns up his coat collar for the deck again; shakes hands all round; and goes laughing out into the weather as to a birthday party.

Where possible, the ship's officers would eat with the first class passengers, and a select few would be invited to dine at the captain's table. Not every captain enjoyed this social side of life but endured it with as good a grace as he could manage. It was part of the job. The officers had other specific responsibilities: one would be responsible for looking after the overall cleanliness, baggage, bullion, navigation instruments and, where appropriate, mail.

For passengers, probably the most important officer aboard was the purser, who was mainly responsible for their welfare and ensuring that the stewards who served them did so well and efficiently. The name is actually a corruption of 'bursar' for it was originally also his responsibility to see that the crew got paid. There is an interesting account of the life of a purser in Michael Brady's blog post *Behind the Purser's Deck: Life at Sea with Paul Motte-Harrison*, 2021. Motte-Harrison worked on P & O liners, running between Britain and Australia. He described his duties, that 'included being able to advise and direct the passengers in any matter relating to their stay in the ship, accommodation, money, postage, passports, landing documents, accounts and entertainment'. There would be a bureau attended by the purser or his assistants where passengers could come for information, but there also a more informal mingling with the passengers at meal times and often friendships were struck – though formalities were preserved at the bureau itself. The purser was also in charge of the stewards, who performed the everyday tasks that visitors would have expected in a first class hotel. One purser wrote a parody of Kipling's famous poem *If.* Here are a few of the verses.

> If you can keep your head when half the boat train
> Lose steamer trunks, and blame the loss on you;
> If you can be mistaken for the boatswain
> And not get rattled at that error, too;
>
> If you can dance with those not young and pretty;
> Stick on the job for twenty-four hours on end;
> And help to organise the Sports Committee,
> Yet still count every passenger your friend.
>
> If you can drink till dawn and not get blotto,
> Avoid the boat deck when the moon is bright,
> And never quite forget that little motto,
> Which runs, 'The customer is always right'.
>
> If you can flirt with dowagers through dinner,
> To doubtful ladies a bit terser,
> Yours is the ship and every soul that's in it.
> And – what is more – you'll be a purser.

For most passengers, their chief contact was with the stewards, who looked after them. Rupert Scott-Padgett made several voyages on P & O liners between Britain and India in the early twentieth century and included this account of their duties:

> The life of a steward aboard one of these ships is not an enviable one. He has to be up at work at four o'clock, washing and scrubbing the saloon, to wait at table four times a day, to make the beds, and attend to the cabins, and to be generally useful among the passengers, rarely finishing before ten o'clock at night.

One vital feature for any ocean liner, was the supply of first class food and drink. In the very early days, livestock had to be kept on board to provide fresh food, but with the arrival of the larger steamships, space could be found for rooms packed with ice to help preserve food for the voyage. The big improvement came in the latter part of the nineteenth century, when refrigerators, powered by the ship's engine room, were installed. There needed to be enormous quantities loaded on board for the liners of the twentieth century. Here are some of the basics for an Atlantic crossing on the *Queen Mary*: 20 tons of beef; 25 tons of potatoes; 4,000lb of tea and coffee; 70,000 eggs; 5 tons of bacon and ham; 9 tons of fish; 160 gallons of olive oil and 500lb of smoked salmon. Liquid refreshment for the bar was equally lavish: 5,000 bottles of spirits; 40,000 bottles of beer; 10,000 bottles of table wine; 60,000 bottles of mineral water and 6,000 gallons of draught ale. No one was likely to go hungry or thirsty.

This mountain of food supplied the galleys that looked after both passengers and crew. A guide of 1910 described the galley of a liner of the period.

> The culinary apparatus used on the modern steamers comprise steam boiling apparatus for boiling vegetables like potatoes, kitchen ranges of the most modern construction, and ample facilities for grilling. The mechanical equipment is very considerable consisting of coffee mills, spice grating machines, cream whipping machines, mayonnaise mixers, knife cleaning and sharpening machines, and buffing heads for polishing silver, as well as the dishwashers. Adjoining the kitchen are the pantries, where the warm beverages are prepared. The ingeniously constructed mechanical apparatus for boiling eggs will also be found which raises the eggs out of the hot water in exactly the number of minutes required.

A separate bakery provided fresh loaves, rolls, cake and pastries every day. Although livestock were no longer needed, there was a large tank of fish kept and one liner had a small market garden on board where fruit such as strawberries were grown.

The finest chefs were employed to provide the luxury menus that first-class passengers expected to receive. According to the guide quoted above, while the first class received the equivalent menus to those of a five-star hotel, the second class were still well provided for, their menus being like those served in decent country hotels. Apart from the passengers, there were also the entire crew to be fed. Altogether, it was an operation that not even the grandest of grand hotels would even contemplate taking on.

One department on board ship that passengers hoped never to need was the surgery with its resident doctor and nursing staff. They had

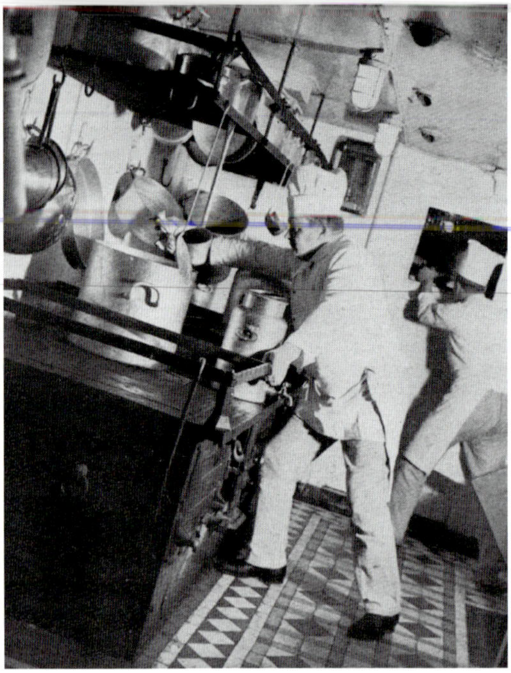

to be ready to cope with any emergency, but inevitably often spent a great deal of time doling out sea sickness pills.

Entertainment on board included regular dances and in the twentieth century the dance band became a regular part of the ship's complement. One of Britain's best known dance bands from the Thirties onwards was Geraldo's, an exotic name adopted by Londoner, Gerald Bright. Passengers in the late Forties may not have known it, but in the band that was playing their strict tempo tunes in the evening might be a few aspiring jazz musicians who would go on to become famous. This was a time when American musicians, such as Charlie Parker and Dizzy Gillespie, were evolving a new kind of jazz that came to be known as bebop. These musicians were unable to come to Britain to play, so the British were happy to sit in the Geraldo orchestra just to have the chance to visit New York clubs such as Birdland. Among those who made the voyages were Johnny (later Sir John) Dankworth, who went on to start his own bebop band, the Johnny Dankworth Seven, and Ronnie Scott, who co-founded London's most famous jazz club. They and others had joined what they came to know as 'Geraldo's Navy'.

This photograph of a ship's galley shows the difficulties the cooks faced when having to work in rough weather.

One group who it is easy to forget, but who were vital to the running of the ship, the ordinary seamen, were scarcely noticed by the passengers but were essential in keeping the ship running.

An unnamed dance band on an ocean liner.

Chapter Seven

Passengers and Life on Board

It is generally assumed that there is something romantic about crossing an ocean in a liner, especially for those fortunate enough to be travelling first class. I had that privilege in the 1950s, when I was called up for National Service in the RAF and was sent for aircrew training in Canada. I returned as Pilot Officer Burton, and being an officer I was, of course, to travel first class, on the Cunard liner MV *Britannic.* One thing you soon discover is that, no matter what class you travel, for most of the voyage you will wake up, look out, and find nothing much to look at outside the ship, apart from a lot of waves. Irving Berlin wrote a song about joining the navy to see the world, with these lines:

And what did we see? We saw the sea
We saw the Atlantic and the Pacific
But the Pacific isn't terrific
And the Atlantic isn't what it's cracked up to be.

Which pretty well sums up the view from a liner's decks. Passengers have to find their own amusements and take advantage of everything

The MV *Britannic* in which the author crossed the Atlantic when a young man.

the ship can offer to help them. Unfortunately, for a 22-year-old, first class was almost entirely occupied by the somewhat elderly – or so they seemed at the time – and all the fun seemed to be happening in second class. But here the rules of the ocean liner come into play – which are basically that ne'er the twain shall meet. But there was one great compensation to keep one going – amazingly good food. It is, I believe, the only time I've actually eaten caviar, and it is unlikely to be repeated. Meals were the highlight of the day, so a good place to start the passenger experience is to look at how the food on voyages changed over the years, and how it varied from class to class.

In the sailing ship years, the diet was limited by availability. With no way of knowing how long a voyage might take, meals could never be lavish – no one wanted to go hungry if the ship was held back by bad weather. For steerage passengers, as explained earlier, the food was often wretched and was handed out at the whim of the captain. When it came to the first steamer crossings, there were notable improvements. Dickens, once he had got over his earlier sea sickness, on his journey to Halifax, Nova Scotia, began to enjoy the food. He made no mention of breakfast but gave details about the rest of the day. It is clear that meals were not simply supplied to satisfy appetites but were also a way to pass the time:

> At one, a bell rings, and the stewardess comes down with a steaming dish of baked potatoes, and another of roasted apples; and plates of pig's face, cold ham, salt beef; or perhaps a smoking mess of rare hot collops. We fall upon these dainties; eat as much as we can (we have great appetites now); and are as long as possible about it.
>
> Meal over, the passengers doze, read or chat among themselves, until dinner time.
>
> At five, another bell rings, and the stewardess reappears with another dish of potatoes – boiled this time – and store of hot meat of various kinds; not forgetting the roast pig to be taken medicinally [it was suggested as an unlikely cure for seasickness]. We sit down at table again (rather more cheerfully than before), prolong the meal with a rather mouldy dessert of apples, grapes, and oranges; and drink our wine and brandy-and-water. The bottles and glasses are still upon the table and are rolling about according to their fancy and the ship's way.

The rolling fruit was not a problem as the tables were fitted with fiddle boards, raised edges that stopped anything sliding off to bounce around the floor. The passengers spent the evening playing whist, with an interruption for tea and toast later in the evening. It can hardly be described as a gourmet's delight.

Steerage passengers in steamers could enjoy far better conditions than they had in the old sailing packets. In 1857, Inman began his own line, specialising in carrying immigrants. First, however, he and his wife crossed in an emigrant ship to see what improvements he could make to attract passengers. For a start, each emigrant had a separate berth, with women berthed separately. One of the problems on emigrant ships had been the spread of infectious diseases. On Inman ships, all would-be passengers were given a medical check before coming on board and if they failed, then they were not allowed to proceed. And there was a doctor on board for the voyage. The other big change was that all were given regular meals. Each passenger had to bring their own cutlery, tin plate and a quart mug. They got three cooked meals a day, but when it came to retiring for the night, they had to bring their own bedding. This was not necessarily due to Inman's benevolent feelings for his fellow creatures: because he was offering shorter crossing times and better conditions on board, he was able to charge more than double the fares asked by the sailing packets. This was still a lot less than cabin passengers paid. An advert for the Richardson line of 1853 offered rates of between 15 and 21 guineas per berth, while steerage went for £8 16s. But there were few cabins and the ship was licensed to carry 645.

By the end of the century and on into the next as liners got bigger and attracted more and more wealthy travellers, the fare on offer changed dramatically. Partly, of course, it came about because refrigeration made it possible to keep food fresh and shorter crossing times put less of a strain on resources. The liners were able to offer menus comparable to those of a five-star hotel to the first-class passengers. There would be a substantial breakfast, an immense lunch and an even more elaborate dinner. Surviving menus from the *Titanic* show the variety of food available. Here are the items on offer for a first-class dinner. For a start there was choice of hors d'oeuvre, consommé or cream of barley soup. After that there was a huge selection – salmon, filet mignon, sauté of chicken, roast lamb, duckling, beef sirloin, roast squab and different vegetables. You could round that off with pâté de foie gras, Waldorf pudding, peaches in chartreuse or ice cream – and, of course, a wide range of wines and spirits were also available. In third class, the choice was rather more limited – so much so that the whole day's meals went on one menu card. Breakfast on offer was, porridge, smoked herring, jacket potato, ham and eggs, bread and butter, marmalade, tea and coffee. Dinner was at mid-day: rice soup, roast beef and gravy, sweet corn, boiled potatoes and plum pudding. Tea was cold meat, cheese, pickle, bread and butter, stewed fig and rice with tea. A late night supper offered gruel or bread and cheese. The second class was, as one

Immigrants crowded onto an Atlantic liner.

might imagine, somewhere between the luxury of first class and the austerity of third.

It was not just the food that made meal times different depending in which class you were travelling; the contrast in the surroundings were even more pronounced, as the photos of the facilities on the *Mauretania* vividly illustrate.

Meals were an important part of the day and for the first-class passengers important social occasions, where everyone dressed appropriately, gowns for the women, full evening dress for the men. In fact, if you were travelling first class you had to have a wardrobe for every occasion, changing throughout the day – and to be fashionable you would not want to be seen in the same old clothes too often. So that demanded a lot of luggage, usually in the form of cabin trunks, which were in effect miniature wardrobes. They could be stood on end and opened up, so that some items could be stored in drawers on one side, garments hung up on the other. The very wealthy would have a

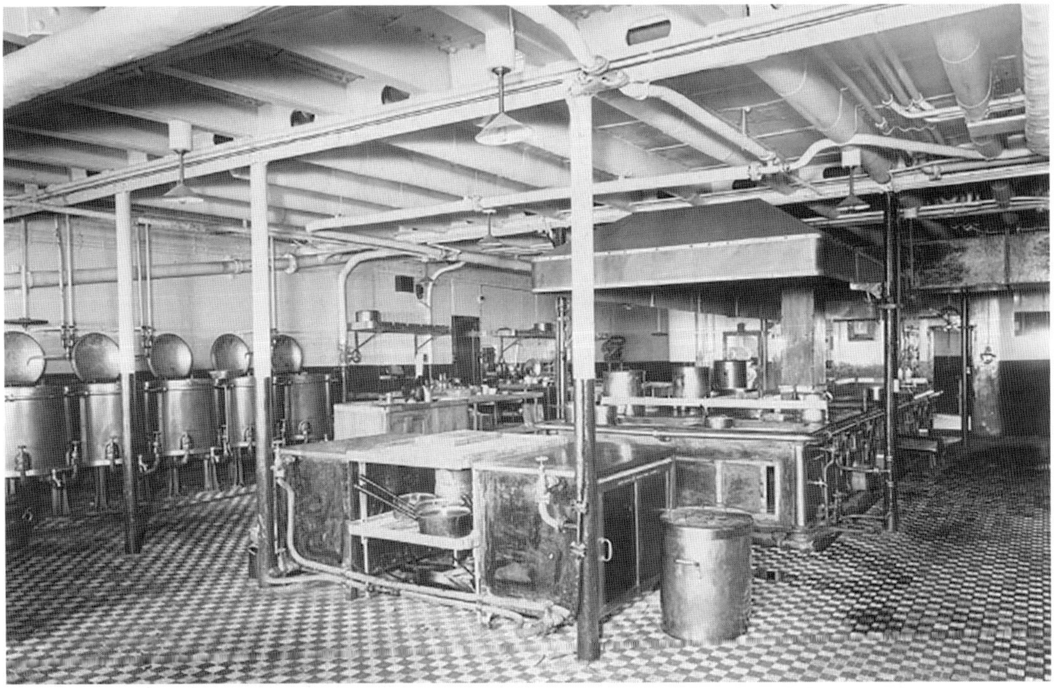

The very modern kitchen of the *Lusitania*, where exotic meals were prepared.

The magnificent first class dining saloon on the *Mauretania*.

The considerably less magnificent third class dining saloon on the *Mauretania*.

whole series of these trunks and other items, some of which would be stored in the hold, but the rest kept for use on the voyage. The Duke and Duchess of Windsor became notorious for the sheer volume of luggage they felt essential to have on board. When it got to the point when they wanted some ninety pieces with them, Cunard hinted that this was perhaps a little extreme. The Duke was deeply offended and never travelled Cunard again.

In the early years, conditions on board were comparatively primitive. Lighting was by oil lamps, but there was an attempt to try something better, when the White Star were building their flagship, *Adriatic*. While she was being fitted out in 1871, they introduced a mechanism for extracting gas from coal that could be used for lighting, just as it already was in buildings on land. Unfortunately, it was soon discovered that gas leaks were a major problem and the whole machinery was removed before launch. The *City of Berlin* had a very small generator installed that lit six lamps, but the idea of lighting an entire ship originated in America. On New Year's Eve 1879, Thomas Edison gave his first public demonstration of electric lighting. Among those who went to see this exciting new development was Henry Villard, President of the Oregon Railroad and Navigation Company. They had a new ship under construction and Villard was keen to install this new system. The ship was being built by Roach & Sons on the Delaware River, and Roach flatly refused to have anything to do with such a revolutionary idea. So, the vessel, the SS *Columbia,* set sail for New York with no lighting of any kind. Once she was berthed, the Edison Company set to work installing a steam generator, and so she became the first ship in the world to have this new, far superior lighting system. She was used mainly on the west coast, between Portland, Oregon and San Francisco. Others soon took up the idea. In 1881, Cunard installed Edison's electric lighting on the SS *Servia* in all the public rooms and in the engine room, but not yet in the individual cabins. Inevitably, all liners would have electric lighting within the next few years.

Lighting was not the only additional comfort that became available to passengers in the late nineteenth century. The White Star Lines' *Britannic* had a huge steam-powered fan that blew cold air round the ship in summer and hot air in winter. Unusually for the time, this was not just for first-class staterooms but was extended to the steerage as well. One innovation was the introduction of flush toilets, no doubt welcomed by passengers, but appreciated even more by the stewards, who no longer had to empty and clean chamber pots. Over the years, as liners got ever larger, more and more facilities were added – many exclusive to first class. A good idea of just what a splendid liner might offer – and who got to use what and when – can be found in the pamphlet handed out to passengers on the *Olympic,* sister ship to the more famous *Titanic.* The dining saloon was exclusive, but there was also a restaurant open from 8am to 11pm, where passengers could book tables. For passengers who had not booked to use the dining saloon and opted to pay as they go, there were special allowances. Next to the restaurant was the Café Parisien – 'with large windows overlooking the sea, is a charmingly decorated café, where light refreshments may be obtained'. The Veranda Café and Palm Court was also available for light refreshments on A deck.

Among the other facilities was a Turkish Bath, 'consisting of the usual steam, hot, temperate, shampooing, and cooling rooms'. Ladies could use the bath in the morning, gentlemen in the afternoon. The swimming pool also had similar restrictions. There was also a rather alarming facility available, called an 'electric bath'. This was a device that was considered very therapeutic. Basically, it consisted of an array of ultra-violet lamps set around the side and on the lid of a cabinet. The user was closed in and the lamps switched on. Usually, the bath attendant would set a timer. There was a small charge for using this contrivance. The more energetic passengers could use a gym 'fully equipped with modern appliances' and there were set times for women and men, and an afternoon spot available for children. A squash

The Veranda Café on the *Mauritius.*

court could be booked. A special section dealt with the 'Lounge and Reception Rooms':

> These rooms are situated on Deck A and at the entrance to the Main Dining Saloon on Deck D respectively. They are intended for the use of both Ladies and Gentlemen, and afternoon tea and after-dinner coffee, while liqueurs, cigars, cigarettes, chocolates, and candies may be purchased there.
>
> A special DANCING FLOOR has been fitted in the Reception on Deck D.
>
> Books may be obtained from the Bookcase in the Lounge on Deck A on application to the Steward in charge.
>
> By special arrangement with 'Harrods' Library, a supply of recent works is placed on board each voyage, as a supplement to the permanent collection of standard works.

Various services were available: a clothes pressing room was 'in charge of an expert attendant' and a stenographer could be hired via the Purser's office. Not everyone got the same facilities; electric heaters

The swimming pool on the Hamburg-American liner *Vaterland*, in what is probably a publicity photograph.

'under control of passenger' were fitted in all the staterooms on A, B, C and D decks but only 1-72 on E. There were three elevators connecting the five decks.

In the late nineteenth century, many liners introduced a new game to help keep passengers amused; deck quoits. It is a very simple game; a target of concentric numbered rings is marked out on the deck. The players have three rope rings to throw, and score according to where it lands in the circle. Shuffleboard was a variation, similar to curling, but on the deck not on ice. It is hardly energetic, but for those who preferred something even quieter, card games such as bridge or whist were often set up. The need to prevent boredom was less essential as crossing time became ever shorter, down to three or four days in the twentieth century. It was a different matter on the long voyages to Asia and Australia. In the early days, these long voyages could become tedious, in spite of stops along the way to take on fuel and provisions. Games were organised of the sort that you might find at a village fete. There might be sack races, egg and spoon or tug of war and a variety of odd games that nowadays have presumably been long forgotten – what, for example, was a tie and cigarette race? There are even reports of cricket matches being played – though hitting a six presumably meant the end of the match as the ball disappeared beneath the waves.

A game of deck quoits in the 1930s.

One curious custom was unique to these voyages – the ceremony of crossing the line. The actual origins of the rituals carried out when crossing the equator are uncertain, but it was already common on British naval vessels by the eighteenth century. It was basically an initiation ceremony for crew members on their first crossing. It was then quite a brutal affair, with the victims, known as 'Pollywogs', being subjected to some incredibly rough treatment, including being thrown overboard on the end of a line and dragged along behind the boat. Needless to say, once the ceremony had been adopted by passenger liners, it had taken on a very different character. Those who took part were brought up before the court of King Neptune and inevitably found guilty and some suitably absurd punishment meted out. It was a chance for everyone, including the ship's officers, to let their hair down, dress up and have a bit of fun. Those who got through the ceremony received a certificate.

Rupert Scott-Padgett was quoted earlier on the work of the stewards and he also gave a full account of his three-week voyage to Bombay on the SS *Ranchi*, calling in at Gibraltar, Marseilles, Port Said and Aden. On arrival at Tilbury, his luggage was taken to his cabin and unpacked by the steward. 'He laid out my dinner jacket and boiled shirt, wrinkled his nose, then took them away for pressing. One does not dress for dinner on the first night aboard. Good form is always observed with P & O.' He also revealed that it was also considered 'bad form' to dress for dinner while going down the Suez Canal, but why that should be is, alas, not mentioned. He went on to describe a typical day on board, that began when the steward arrived at 7am with a tray of juice, coffee and biscuits. 'Children's breakfast is served at 7.30 and the nannies are there to help the children. They cut their meat and wipe their faces – and keep the peace between the charges while mothers leisurely dress for breakfast at a later time.'

His days were dominated by meals – breakfast at 8.30, coffee at 10, lunch at 12.30, tea at 4, dinner at 7 and a buffet at 10:

> When we are not eating, we passengers spend our time sitting on deck, talking, reading, playing cards, etc ... In fine weather, I can be found sitting on deck, breathing the salt sea air, with my head inside a good book. And if the weather is inclement, I normally withdraw to a corner of the smoking room or writing room. One enjoys a sense of tranquillity in the writing room.

There were other activities organised by the Sports Committee. 'This particularly British institution manages to form itself, voyage after voyage, the ingredients are always there: the born organiser, the games enthusiast, the public-spirited and the energetic and, of course, the busy-body and the exhibitionist who must always be in on everything.' He was clearly not a great enthusiast for energetic activities. 'After lunch and a glass of Allsopps lager, I generally retire to my cabin for a relaxing siesta – what I call my "daily correctly structured and restorative sleep episode".' At the ports where they stopped along the way, it was possible to go ashore, and even spend a night in a local hotel, or go on excursions – a popular one being a visit to see the pyramids.

One of the highlights of the voyage was the Arabian Nights fancy dress party held on deck:

> Above us the night sky was a mass of stars, while the moon bathed the faces of the partygoers with its eerie light. Some of my fellows had gone to enormous length with their attire - eastern potentates and caftan-wearing colonial officers, their faces flushed with wine and excitement – seeming to outnumber those in more modest Bedouin costume, purchased from the bazaars of Aden. Some of the younger memsahibs had entered into the spirit of the occasion, with gaily coloured sandals and saris, while the older ladies looked on with something approaching disapproval.

He eventually got to bed after midnight – and recorded seeing several passengers the next morning looking somewhat pale and unsteady. It had obviously been a good party. Soon afterwards, they reached Bombay. His account makes the whole voyage seem very pleasant, but he is, of course, only seeing it from the point of view of a first class passenger.

The arrival of ship's radio meant that passengers were kept in touch with the latest news while they were at sea, with the ship getting the news over the air and then producing their own newspapers. The passenger on a liner was no longer in their own little community, altogether isolated from the rest of the world. That was rather more

The state rooms of an unnamed ocean liner.

important for those making the long journey to the Far East and Australia than to those making the comparatively short Atlantic crossing.

Inevitably, whatever amusements a passenger might find on board, a great deal of the time would be spent in the cabin, and that could be palatial or cramped, depending on how much you were prepared, or able, to pay. The grandest of all were the suites, with separate bedroom and comfortable sitting area, while at the opposite end of the luxury scale were third class cabins, which had little more to offer than bunk beds, and which would normally be shared. They were, at least, a great improvement over the earlier steerage accommodation, where the passengers were simply herded together in open dormitories. For those lucky enough to occupy the former, they could happily spend comfortable time there, just relaxing or reading one of the books borrowed from the library. They would also have had the spacious promenade decks to wander around and meet other passengers. In

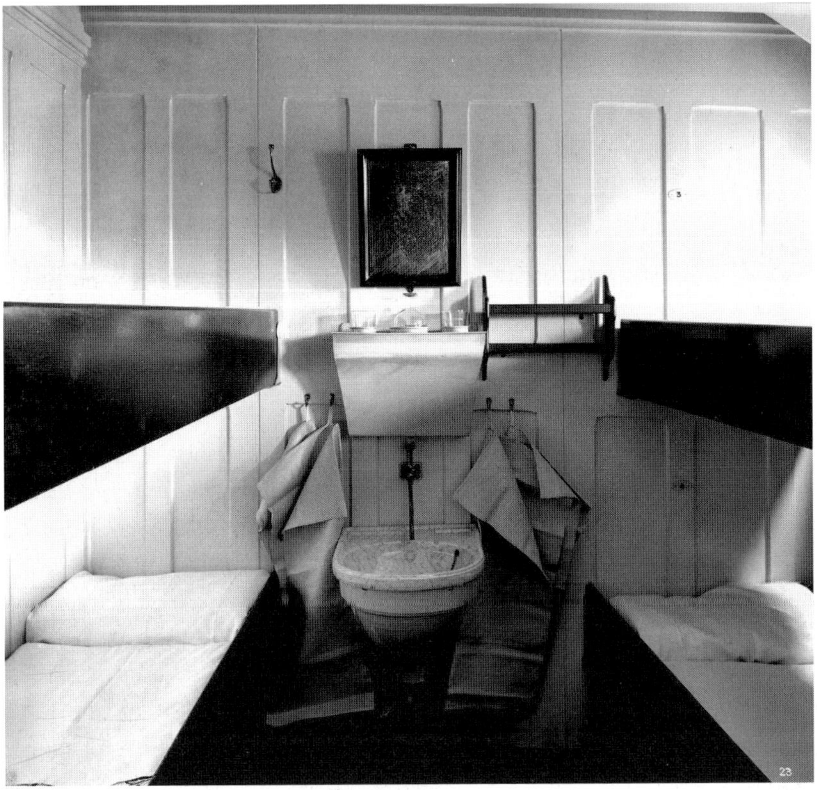

This spartan third class cabin on the *Balmoral* offers very little in the way of privacy.

A two-berth cabin on the American liner *Monterey* that went into service on Pacific routes in the 1930s.

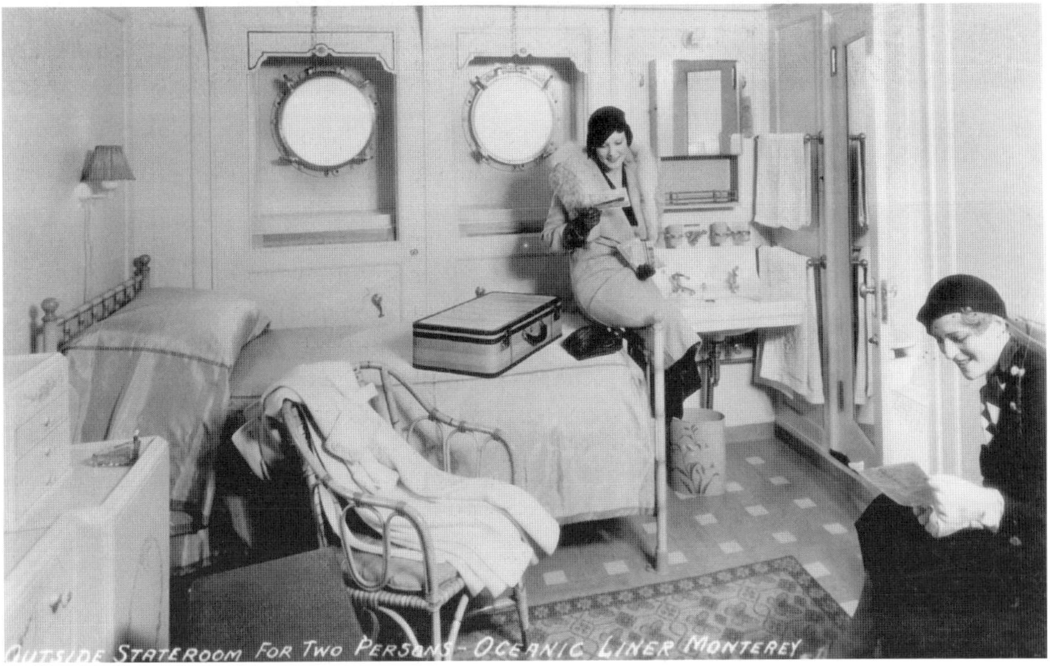

1885, an American, John Cham, invented a folding chair that could be stowed away when not in use and brought out on deck in good weather – hence the deck chair. For those in the third class, there was little option other than to wander round the ship in whatever space had not been reserved for the other classes.

The promenade deck of the *Kaiserin Maria Theresia.*

Steerage passengers huddled together in the small amount of open space available to them on deck, a notable contrast to the scene in the above illustration.

Chapter Eight

The Blue Riband of the Atlantic

The blue riband is awarded to the fastest vessel making a westbound Atlantic crossing. Not all lines followed the same route – some sailed direct from Liverpool, others made their last call at Queenstown in Ireland – so the prize was not for the shortest crossing in days and hours, but based on the average speed for the whole voyage. Brunel's mighty *Great Eastern* had proved disappointingly slow, so in 1845 the honours went to Cunard's *Asia* with an eight and a half day crossing at an average speed of 12.25 knots. The record was soon beaten when the Collins line brought in their successful, but short-lived steamer fleet and held the record from 1850-4. When that company went bankrupt, the American challenge was at an end and it was to be many years before that country competed for the riband.

Competition for many years was between three British companies, Cunard, White Star and Inman, who played a game of nautical leapfrog, passing each other for the prize at regular intervals, while all the time there was a steady increase in speeds, and although in 1856 the record

The Cunard ship *Asia* from the *Illustrated London News*, May 1850.

speed of 13.11 knots was set by the Cunard paddle steamer *Persia,* the future lay with propellor driven vessels in the latter part of the century. Then, in 1882, a new company appeared on the scene – Guion.

The company has a complex history. It began in America as the Black Star line, carrying steerage passengers between Liverpool and New York. Stephen Guion was one of the original partners who came to Liverpool to act as the company's agent. In 1866, they incorporated the ill-fated Liverpool and Great Western Steamship Company into their organisation. Although it was originally based in New York and was mainly financed by American shareholders, Guion registered as a British company. There was a very good reason for this. They had decided to concentrate on providing fast ships for the Atlantic run, but if they had registered in America, they would have been forced to use American built vessels. As they needed modern, iron hulled, screw steamers to achieve their aims and as there was no one in America able to build them, they had no choice. They would be a British company running British built ships, specifically designed to win the blue riband.

Their first efforts were dismal failures. The *Montana* and the *Dakota* were designed for speeds of 17 knots but never managed more than a fraction above 11 knots. They both went into service in 1877 but were both lost at sea off Anglesey in 1880. It was time for a new approach. William Pearce of the John Elder shipyard on the Clyde suggested that what they needed was a powerful ship that would only carry light freight and would make its appeal by offering fast passages. Cunard had already turned down the idea, but Guion accepted and were offered a good financial deal. Altogether, three vessels were built, the *Arizona, Alaska* and *Oregon.* They were successful; *Alaska* held the blue riband from 1882-3, after which it was taken by *Oregon* that held the award until 1885. Guion had originally hoped for a 17 knot ship, but *Oregon* surpassed that, recording a fastest run at 18.56 knots. The line's success was, however, short lived as the Guion family were hit by a series of financial disasters and the company was liquidated in 1885. It was reorganised under Pearce, but the record breaking days were over; Cunard was about to take over the trophy again, ready to do battle with Inman and White Star.

Cunard's *Etruria* held the riband until 1889, when Inman took over with a ship that contained many new and important features, the *City of Paris,* the second liner to bear that name. The Inman line had been made bankrupt, and was taken over by the major creditor, the International Navigation Company, based in Philadelphia. Under the new management, the decision was taken to provide $2million to finance the construction of a ship that would be designed to compete for the blue riband. An earlier design in the Inman era had resulted in

The Guion line's *Alaska* photographed near New York by James S. Johnston, and published by the Detroit Publishing Co.

the *City of Rome* being returned to the builders when she failed to live up to the required standards. It was eventually agreed with the J. & G. Thomson shipyard on the Clyde – a yard later to become famous when it changed hands and became the John Brown yard. The company had started as engine builders and only later turned to shipbuilding. The 1880s were not a good period for Scottish shipbuilders, so they were delighted to receive an order for two ships at $1.85million each. The designs they produced contained a number of very different factors.

The first and most obvious difference was that the two vessels were to be constructed of steel, rather than iron. Generally, the hulls of liners were built on a 10 to 1 ratio, but Thomson wanted a more rigid structure that would be less liable to be troubled by vibrations from the engines, so they built to a ratio of 8.3 to 1. Safety was ensured by the use of a double bottom, as already seen on the *Great Eastern*, and fifteen bulkheads that reached from the bottom of the hull to saloon deck level. The real change, however, was down in the engine room. Here were two identical, triple expansion steam engines, driving two shafts – the first liner to be equipped with twin propellors. The second of the two vessels, *City of Paris*, was launched in 1888 and was a supremely elegant vessel; with her clipper bow and three raked funnels, she seems almost slight compared with other liners of the period, but she was a big ship, with a crew of 362 and able to carry 1,740 passengers. And she did exactly what she was built to do when in 1889 she took the record and at the same time became the first liner to make the crossing at an average speed of over 20 knots. She was not, however, to have it all her own way. White Star were to make their own challenge, with an order for two new vessels, *Majestic* and *Teutonic*. This was not a purely commercial deal, as part of the money was put up by the Admiralty on the condition that the vessels would also be placed on the naval reserve, when they could be converted to armed merchant cruisers. They were built at the Belfast yard of Harland & Wolff and, like *City of Paris*, were fitted with twin screws. In 1891 both ships achieved the fastest speeds on the Atlantic but did not hold the record for long. The following year, *City of Paris* reclaimed the prize.

Cunard also had two new vessels on order, *Luciana* and *Campania*, and to ensure that the vessels also met naval standards the order went to Britain's biggest builder of warships, Fairfield, based at Govan on the Clyde. Again, both vessels were to be driven by two screws, powered by the biggest engines ever put in a liner. The quintuple expansion engines rose to a height of 47 feet, each with five cylinders, two high pressure, one intermediate and two low pressure. With possible wartime use in mind, each engine was in a separate watertight compartment, so that if one was put out of action, the other could continue working. The vessels

City of *Paris*, another photograph taken near New York by James S. Johnston.

were not only notable for their engineering excellence, but were among the most opulent afloat and epitomised Victorian over-the-top design.

Campania was the first to take to the sea in April 1893, but there were a few teething troubles, especially with vibrations. Modifications were made to the *Luciana* that was still under construction, and later to the *Campania* as well. Between them they held the riband from 1894 to 1898, when a new contender appeared in the field, not from Britain or America but from Germany.

The Norddeutscher Lloyd (NDL) had previously been a comparatively minor player in the Atlantic competition. But Germany had changed dramatically in the latter part of the nineteenth century. What had once been a loose federation of independent states had become the German Empire in 1871 and the new emperor Wilhelm II was anxious to match British achievements. He had been present at the review of the fleet of 1889 to celebrate his grandmother Queen Victoria's jubilee and wanted to be able to match this one day with a review of his own mighty navy that would be a match for anything produced by the British. In particular, he wanted Germany to have a magnificent liner that could be converted, as the Cunarders could, into battleships. NDL began work on a splendid vessel, the *Kaiser Wilhelm der Grosse* that was launched in 1897, the first vessel that actually matched the *Great Eastern* for size, launched 37 years earlier. The deck was specially strengthened, so that if needed she could mount eight 15cm guns, four 12cm guns and fourteen machine guns. Passengers were not aware of the military aspect of the great ship, but what they saw impressed them enormously from the start – for here was the first ship to boast four funnels. Inside was equally impressive, from the magnificent first-class dining room that could seat 800 passengers to the smoking room fitted out like a typical German tavern. She was also the first ship in the world to install Marconi's newly invented electric telegraph. Otherwise, in spite of her four funnels, she was still powered by just two triple expansion engines, driving twin propellers. In 1898 she did what she was designed to do – she took the riband away from Cunard with a speed of 21.81 knots, more than a knot faster than the Cunard record. She marked the start of a period of German dominance, with first the *Hapag Deutschland,* then the *Kronprinz Wilhelm* taking the honours, the latter at over 23 knots in 1902. The British government was far from pleased with this new threat to their dominance on the Atlantic and in 1903 they entered into an agreement with Cunard to build two ships to take back the riband. But before that happened, a new technology had already appeared that was to have an immediate impact on the shipbuilding industry.

***Wilhelm der** Grosse* from Cassier's Magazine of January 1898.

Charles Parsons was the junior partner in an engineering firm, Clarke Chapman & Co. They built electrical generators, and instead of using the reciprocating engines, with pistons going up and down in cylinders, they built steam turbines. Here the steam was used to turn blades fastened to a central shaft. Parsons realised that this would be an efficient system for turning a propellor shaft and in 1894 he formed the Marine Turbine Company. They set out to build an experimental craft, appropriately named *Turbinia*. She was a sleek speedster 100ft long but just 9ft beam and certainly looked the part. Her performance on trials, however, failed to live up to expectations, only reaching 20 knots. It was thought this was due to the cavitation effect, that slowed the propeller. Parsons carried out tests using a tunnel and strobe lighting, then took a series of photos that showed the supposition was true. He at once set about changing the design of *Turbinia,* with a compound turbine, driving three shafts.

The redesigned *Turbinia* was an instant success. In 1897, she was demonstrated at the Spithead Review of the fleet, where she dashed across in front of the great line of battleships at 34 knots. A story grew up that the Admiralty had refused to show any interest in the vessel and Parsons had cheekily made the famous run unannounced. A good story – but alas not true, as the vessel had actually been formally invited to take part. Parsons realised that speed itself was not enough; smooth running and efficiency were equally important. He went on to design a steam launch in which 20,000rpm turbine speed was reduced through gearing to 1,400rpm. It was ideal for a new type of naval vessel, the fast torpedo boat, but with further reductions the turbine could be adapted for use in ocean liners.

In 1903, the government entered into an agreement with Cunard for the construction of two ships 'capable of maintaining across the Atlantic a minimum average speed of 24 to 25 knots … in moderate weather'. Cunard were given a loan of £2.6million at what was then a very generous interest rate of 2¾ per cent and a subsidy of £150,000 to cover running costs. The *Mauretania* was to be built by the Swan Hunter shipyard on the Tyne and her sister ship, *Lusitania,* by John Brown on the Clyde. Very exacting safety standards were required. Stress was measured in terms of a wave size to the length of the ship. For these two vessels that meant the wave they had to safely overcome would have to be 760ft long with a 35ft high crest. They were unlikely

The pioneering steam turbine yacht *Turbinia* at speed in 1898.

to meet such a monster, but the safety factor was 3.5 times that. New materials were used – on the *Mauretania,* the midship section was strengthened with high tensile steel. Previously only mild steel had been used in construction, but the new material being much stronger meant that less metal had to be used in construction. The important factor, however, was the change from reciprocating engines to turbines. There were four turbines to move her forward and two more to move her astern. Turbo-generators also provided enough electric power to light a town of 100,000 people. Both vessels entered service in 1907, when *Lusitania* promptly took the lead with a speed of 25.65 knots but two years later *Mauretania* took over with just over 26 knots, a record that was held until 1929.

NDL came back into the contest with two new vessels, *Bremen* and *Europa.* They too had made the change to steam turbines, with four sets driving four propellers, and also had new technological features on offer. One was the bulbous nose, a device which looks much as its name suggests, and was attached to the bows below the waterline, with the object of smoothing out the flow of water over the hull, so reducing drag and increasing speed. It was not entirely new, having already been in use on cargo vessels, but it was the first time it had appeared on a liner. The other novel feature that attracted a great deal of attention was a catapult on the upper deck, for launching a small seaplane to carry mail. Once it was within flying distance of port, the plane was launched, arriving at the destination port long before the liner steamed into view. Everything was outstandingly modern. The turbines were oil fired using a boiler system designed by Professor Gustav Bauer, that used eight steam powered turbines to blow air into the oil burners that heated the boilers. There were four airtight

***Mauretania* to** the right of the picture with *Aquitania* behind her, photographed in the 1920s by Francis Stuart.

boiler rooms that could only be entered through airlock doors. Also on board were four diesel generators supplying electricity for 420 electric motors and everything from the 21,000 lamps on board to the electric lifts connecting the decks. With such an array of power plants, it is no surprise to find the ship required an engineering staff of 170 to keep it all going. The two vessels shared the honours of holding the blue riband, with speeds of almost 28 knots from 1929 to 1933.

Italy decided to join the competition for the blue riband, with two new ships, *Rex* and *Conte di Savoia*. There was nothing technologically novel. Like the earlier holders of the honours, power came from four steam turbines and like *Bremen* they had a bulbous nose at the bow. They were built to carry over 2,000 passengers in four classes, with a new 'tourist class' between second and third. What was new was the modern publicity drive to attract passengers – these were not just a means of getting between Europe and America but were dubbed as Rivieras afloat. The outdoor swimming pool had sand round the edges and large colourful umbrellas so that passengers might think they actually were lounging by the sea at Nice or Cannes – provided they ignored the fact that they were also going up and down. *Rex* held the riband from 1933-5 at a speed of 28.92 knots.

The next contender came from France and is often regarded as epitomising the best of twentieth-century liner design – the *Normandie*. Her sleek hull and clipper bow gave her a very elegant appearance – and the hull itself was designed not only to look good but to make it as efficient as possible. Power still came from four steam turbines that were used as generators to provide electricity for the four electric drive engines. The power output was 133,400KW and when the smaller amount produced by auxiliary generators was added in it was sufficient, according to a contemporary account in *Shipping Wonders of the World*, to power the whole London Underground system. This was all very impressive, but it was the interior and fittings that caught the public's attention.

There was good reason for this, and its origins lay in two American laws, limiting immigration. The first of 1921 put strict limits on immigrants from Europe, and the second of 1924 was even stricter and

An advert for the two Italian Line ships, *Rex* and *Conte di Savoia*. It is significant that the advert is aimed at increasingly important tourist class passengers.

was particularly aimed at anyone of Asian origin. Steerage passengers had made up a large and profitable source of income for the Atlantic liners, but now it was dribbling away. As a result, the balance of the ship's accommodation shifted. The *Normandie* had space for 848 first class passengers, 870 tourist class and just 456 third class. The fourth class that had been available on ships like the *Bremen* had ceased to exist. The interior has been described already in Chapter 4 but there were several very novel features as well. The ship contained twelve decks with ample space for a whole variety of specialist rooms from chapel to cinema. Although she appeared to have three funnels, one of them was a fake and housed air-conditioning machinery and kennels. This left a lot of space on the top deck for passengers – and that included a full-sized tennis court. Even the third-class had their own sun deck in the stern. She won the riband on her maiden voyage in 1935 and held it from 1935-7, almost topping the 30 knot mark, but just stopping short at a recorded average speed of 29.98 – that 0.02 must have infuriated the proud owners. She did, however, have her own financial problems – the Wall Street Crash of 1929 had sent the western world into a period of acute financial depression, and this meant fewer customers for luxury travel when *Normandie* was launched three years later to the day.

Britain was also in the business of trying to build a prestige record-breaking liner. In 1930, the John Brown yard won the contract for the new ship that specified that 'the whole of the vessel's hull, machinery and equipment is to be of British manufacture'. Although the main work would be Brown's, there were over seventy sub-contracts, bid for

***Normandie* at** Le Havre; the after funnel, from which no smoke is emerging, is a dummy.

by 100 companies. She was designed for speeds in excess of 30 knots, with power from four steam turbines. Work got under way, but as the world financial crisis hit, the money ran out and in December 1931 with some £2million already spent, work came to a halt. Three thousand men were laid off. The great unfinished hull sat there until 1934, when a government grant enabled the work to be completed. Cunard were keen to get as much publicity as possible for the project and produced a booklet that was sent to Sir Thomas Bell of Brown's, which got a frosty response; 'If we cannot build ships better than Cunard can produce booklets – Gawd help us!' A suggestion that the BBC might come and make programmes at the yard was firmly rejected – 'it would be prejudicial to the discipline'. Brown's did, however, seem happy to discuss whether the invitations to the launch should specify 'Undress Uniform' or 'Morning Dress', such was the old fashioned, rigid formality of the day. But there was nothing old fashioned about ship No. 534 as she was still known. She incorporated all the latest design features, and she was finally launched in September 1934, and named *Queen Mary*. She finally won the blue riband in 1936, the first liner to have an average speed of over 30 knots.

The *Queen* *Mary* at speed.

During the Second World War, records for Atlantic crossings became totally irrelevant, and afterwards there was to be a period of recovery, before anyone began to contemplate building new luxury liners. During the war, liners had played an important role as troop carriers and afterwards, when the United States line decided to build a new ship, they went into partnership with the government, but with the proviso that construction had to allow for possible later use as a troop ship. Construction costs were put at $78million, of which the government underwrote $50million. The most novel feature of the new

It is often difficult to get an idea of the scale of the machinery that drove ocean liners: this photo shows men preparing just one element of a steam turbine for the *Queen Mary*.

ship, the *United States*, was the superstructure, constructed entirely from aluminium, with a huge saving in weight overall. The method of construction was also new – she was built in a dry dock, with prefabricated sections being brought to the site and assembled. On her maiden voyage in 1952, she easily broke the old *Queen Mary* record with a speed of 34.51 knots. It was a record destined never to be broken, and although faster vessels would cross the Atlantic over the years, they were not the ocean liners of old. Their days were coming to an end.

So far, most of the emphasis has been on the Atlantic liners, but from the early nineteenth century onwards there was an equally important development in vessels carrying passengers to the Far East and Australia.

The *United States*: last holder of the Blue Riband.

Chapter Nine

The Far East and Beyond

Sailings to the Far East became ever more important as Britain's relationship with India changed. Instead of being foreigners who came to trade, they had become the masters who ruled. There was no immigrant trade as such from India to Britain, but there was a need for a regular service for what was largely a male clientele, travelling to serve at first in the East India Company and later in the Indian Civil Service. Few were married when they left Britain, so there was also a number of young, single women who took work in India or travelled there for interest. As they were also said to have had eyes open for attractive and eligible young men, they became known as the 'fishing fleet'. Most of those who travelled to Australia in the early years did so reluctantly, having been sentenced to transportation, often for what would now be considered minor offences scarcely even warranting a jail sentence – one example is a poacher who was convicted of taking a rabbit from a warren on private land. Political offences could also lead to transportation, the most famous example being the Dorset labourers, who in 1834 decided to form a trade union to fight against cuts in their basic wage. This was not in itself illegal, but they swore an oath of secrecy, and that was. They were transported and became famous as the Tolpuddle Martyrs. The first convict fleet left for Australia in 1787 and by 1860, 162,000 men, women and children had been sent as punishment to Australia. Many were for terms of a few years but having endured the miseries of the crossing from Britain, few chose to return home at the end of their sentences, even if they had the money for the fare. It would be years after transportation ended before more conventional passenger traffic was established on a large scale.

Regular service to the Far East developed in stages. The most famous company to offer the service began with far shorter voyages. In 1815, Brodie McGhie Willcox, who had been born in Belgium in 1786 to an English father and a Scots mother, began a shipbroking business in London. He was later joined by Arthur Anderson, a Shetlander who had spent time at sea and was initially taken on as a clerk, but later became a partner. They began establishing routes to Spain and Portugal and during the Portuguese Civil War of the 1820s used a small schooner to take arms for the royalist faction. Anderson himself went with the first shipment, travelling under the unimaginative pseudonym 'Mr. Smith' to ensure the money was there for payment. They were active again during the first Carlist War of the 1830s, again supporting the royalist against the liberalist side. By now, sail was already giving way to steam and they chartered vessels from the Dublin and London Steam Packet Co. The two partners of that company, Captain Bourne and James Allen company met up with Willcox and Anderson, and together they decided to collaborate in running regular steamer services to Spain and Portugal and set up a new company, the Peninsula Steam Navigation Company.

The new company's first ship, *Iberia*, was built in 1836, a wooden hulled paddle steamer. She had accommodation for 37 first-class passengers and 16 second-class, with a crew of forty-one. She was awarded the mail contract and offered weekly sailings, starting in London and calling in at Falmouth, Vigo, Oporto, Lisbon and Cadiz, before ending at Gibraltar. Connections were then possible for Malta and Corfu, and once a month there was a further connection to Alexandria. It was a modest beginning, but the company soon had an opportunity to expand. In 1839, the Governor General of India, Lord William Bentinck, invited tenders for a regular service from Britain. Willcox and Anderson won the contract with a bid of £34,200. They were already aware that they would need a minimum of two ships for the job, as some of the passengers to Alexandria were already using the port as a stopping off point for India, travelling overland to resume the ocean voyage. Given the extended service now being offered, the name was changed to Peninsula and Orient, or simply P & O.

Two new steamers were required. The first was to take passengers from Southampton to Alexandria. From there, passengers for India would travel down the Nile then overland by mule train to Suez on the Red Sea for the rest of the journey. This part of the journey was made considerably easier in 1858, when a railway was opened to link the two ports, reducing a journey that was scheduled to take 88 hours to a single day. The first of the new ships to be delivered was the *Hindostan* that covered the second part of the journey from the Red Sea to Calcutta (Kolkata), calling in at Colombo and Madras en route. Like *Iberia*, she was a wooden paddle steamer, but considerably larger, able to take 102 first-class and 50 second, with a crew of fifty-three. Her maiden voyage was in 1842, and she continued in service until 1864, when she sank in a cyclone off Calcutta. The journey down the Red Sea was notorious for being uncomfortably hot, and the best cabins were those that did not get the sun beating down on them all day: so, the rule was port out, starboard home – POSH. If you could afford to pay the extra for the privilege, you were a posh person – though the Oxford English Dictionary does not accept this version, simply listing the origin of the word as 'obscure'.

The *Chusan*, the first steamer to arrive in Australia, being welcomed into Sydney harbour.

A rare photograph of the deck of the *Chusan*.

In 1844, P & O were invited to tender for an extension of their existing routes to Penang, Singapore and Hong Kong, and the first vessel reached Singapore the following year. The next advance was to extend even further, from Singapore to Australia. The first vessel to be built for the new service was the *Chusan*. Technology had now moved forward, and she was an iron hulled vessel with a single screw propeller. She left England on 15 May 1852 and arrived in Sydney to an enthusiastic reception on 3 August. The enthusiasm is understandable; previous voyages under sail had taken as much as 130 days. She was not, however, the first steamer to arrive in Sydney from Britain. That honour goes to the paddle steamer *Sophia Jane*, but her paddle wheels remained stowed away on board for the whole journey and were only put to use after she reached Australia in 1831. There was one feature of the ship that one would not expect to find on an ocean liner; she was heavily armed. She carried one 32-pounder gun, one 18-pounder and four 12-pounders. There was good reason for this. The area round what is now Malaysia and Indonesia was notorious for piracy. It was to remain a serious threat to shipping for much of the nineteenth century. There was, however, another aspect to the P & O's trade at this time that hardly gave them a claim to a much higher morality than the pirates; one of their most lucrative cargoes was opium.

Details of the voyage were recorded in the *Sydney Morning Herald*. They gave notes on the accommodation on board: 'The cabins are very lofty – ten foot six inches, which of course makes them very airy. Her state rooms are fitted out for three passengers each, and have all the modern improvements and conveniences.' They hit several spells of severe weather, and on one occasion the captain was forced to heave to for twenty-four hours to ride out the storm. Then, when a gale hit later in the voyage, a sailor manning the main topsail was blown overboard. In spite of the high seas, a boat was lowered, but the crew struggled to save themselves, and it was two hours before they could be got safely back on board. The man who fell was never recovered.

Voyages to the Far East and Australia were revolutionised by the opening of the Suez Canal in 1869. Passengers no longer had to use the overground route between Alexandria and Suez, which was good news for them, but not quite so good for P & O. They had invested heavily in the overground facilities, and had ships designed specifically for the two different sections. They also had a mail contract that still required them to send Egyptian mail overland. A new managing director, Thomas Sutherland, took over in 1872 and at once put in an order for seven new ships.

They were comfortable vessels, but had no particularly novel features, and served the company well. One, however, was destined to have a

An unnamed Dutch steamer about to pass a dredger in the Suez Canal.

more dramatic history than the rest. The *Bokhara* was a single screw steamer designed to take 133 first class and 56 second class passengers. She set off on her maiden voyage from Southampton to Bombay on 10 April 1873, and then travelled on eastward. On 21 October, she hit a rock while leaving Hong Kong, was temporarily beached at Kowloon and then brought back to Hong Kong's dry dock for repairs. But it was in 1892 that tragedy struck. On 8 October, she left Hong Kong for Shanghai with 141 persons on board and a valuable cargo of specie valued at $200,000, roughly $700 million today, together with 1,300 bales of silk and 800 tons of tea. En route to Shanghai they were hit by a typhoon in the Taiwan Strait, and the heavy seas swamped the boiler room, leaving the ship helpless in the storm. The captain attempted to head for nearby land, but before she could reach safety, she hit reefs, which tore the hull open. Only 25 of the 148 on board survived. They were able to shelter in a hut on Sand Island, from which they were taken by Chinese fishermen and eventually picked up by a Douglas Line steamer.

P & O expanded over the years, not always successfully. Early attempts to start a service to Australia via Cape Town failed due to competition from the Union-Castle Line, but re-emerged in 1896, when they combined with the Blue Anchor Line and began a regular immigrant service to

Australia. More mergers followed, notably with British India, making it the largest fleet of liners in the world. There was a massive rebuilding programme needed in the years following the end of the First World War. Work had actually started on the company's first three funnel ship, the *Naldera,* in 1913, but she was only completed in 1918 and finally entered service in 1920 on a maiden voyage to Sydney. She had a quadruple expansion steam engine, driving twin propellers that gave her a maximum speed of just over 18 knots. She was stood down in 1938, but almost played a role in world history. She was to have been used to carry the British Legion Volunteer Police, who were to supervise voting in Czechoslovakia where a referendum was to be held on Hitler's plans for partitioning the country. But the British government's appeasement policy virtually handed the country over to Hitler, so there was no vote to oversee, simply a takeover to accept.

One passenger who recorded life on board a P & O ship in 1929 in his book *When the Going Was Good* was the novelist Evelyn Waugh. He was only on board for a short time – joining the *Ranchi* in Egypt and leaving again at Malta. He described a chaotic scene, when the vessel berthed at around midnight on a Sunday evening and only remained in port for two hours:

> During those two hours the town, which, as usual, was feeling the ill-effects of its Saturday night at the Casino, suddenly woke again into life. Simon Arzst's store opened; the cafés turned on their lights

P & O liner RMS *Maloja* approaching Sydney in the 1930s.

> and dusted the tables; out came the boot-cleaners and postcard sellers; the passengers who had stayed on board through the canal came ashore and drove around in two-horse carriages; those who had left the ship at Aden for a few hours at Cairo, and had spent all that afternoon on the quay in a fever of apprehension that they might miss her, scuttled on board to their cabins; half the residents of Port Said had business of some kind to transact on board. I went down to the harbour that was like noon in the City of London.

The ship was comparatively new, having only been built in 1925. Waugh had booked to travel second class, which meant sharing a cabin with a stranger. He had a good night, but did not enjoy the voyage. Most of the passengers were soldiers, with their wives and families coming back to Britain from India. He found his fellow passengers pleasant but there were, he wrote, 'so many of them':

> The lounge and smoking-room were comfortable and clean and well ventilated and prettily decorated and all that, but they were always completely full. On the decks there were no deck-chairs except those the passengers provided for themselves; the three or four public seats were invariably occupied by mothers doing frightful things to their babied with jars of Vaseline. It was not even possible to walk round with any comfort so confined and crowded was the single promenade deck.

***Ranchi*, the** P & O liner on which Evelyn Waugh travelled in the 1920s.

He noted that the officers 'cultivated an extreme freedom of dress' during the day, unlike the first class 'in white flannels and parti-coloured brown and white shoes'. Come the evening, however, the officers appeared for dinner in correct evening dress; some standards obviously had to be observed. There seems to have been little in the way of entertainment, apart from a band that appeared for an hour from the first class to play selections from Gilbert and Sullivan.

The outstanding vessel of the interwar years has to be the *Viceroy of India*, launched in 1928, if only for the technology, as the first big liner to use the steam turbine as a generator for an electric motor. She was a handsome vessel, and seen out on the water, the very picture of modernity. The interior, however, like so many other liners of the period was firmly tied to the past, with public rooms that wanted to emulate some grand country mansion, including a baronial hall complete with massive stone fireplace. Everything was aimed at the luxury market, for the liner depended on it, with accommodation for 415 first-class passengers but only 258 second-class, which explains why of the 413 crew normally on board, 248 were pursers and stewards. Things were very different following the end of the Second World War. Now the main passenger trade was not in the wealthy, but with the 'Ten Pound Poms'. This was a scheme introduced in 1945, in which the Australian government subsidised the fare to Australia, which for tourist class was £120 – so that the British emigrants were only required to pay £10 each. By 1972, when the scheme ended, over

A 'Strath' class P & O liner approaching Bombay.

one million people had set off for Australia and an uncertain future. There were several stipulations – they had to be under 45, pass a medical and stay for at least two years – if they left before that, they had to pay their own fares home. Of those who took the ten pound trip, about a quarter returned to Britain, much to the scorn of those who remained – from being ten pound poms, they had become whingeing poms.

A number of vessels were used, including P & O's latest *Canberra,* that came into service in 1961. She looked quite unlike any of the earlier P & O ships, with her twin funnels set aft instead of midships. In contrast with the passenger accommodation on *Viceroy of India, Canberra* had 548 first class and 1,690 tourist class places. This scheme is very well known, but an earlier scheme has received less publicity. From the 1860s onwards, there had been a scheme for sending orphaned children to Australia. One of those, who had come from an orphanage, looked back on his trip to Australia in the 1920s with nostalgia: 'One thing that amazed me was that we were treated the same as everybody else on board by the passengers and crew. We were treated as human beings.' That is a real tribute to the P & O crew of those days.

P & O's rival on the South African route was the Union-Castle Line, which began as two quite separate companies. The Union line began modestly in 1853 as the Southampton Steamship Company,

A P & O promotional photo of a bar on board, emphasising the relaxed atmosphere on board in post-Second World War years.

Canberra **at** Sydney, passing the Sydney ferry.

with the less than romantic job of bringing coal from South Wales to Southampton specifically to bunker P & O ships. They had just three small vessels, the largest only 530 tons. Three years later, they decided to expand to become general freight carriers, and the name was changed to the Union Steamship Company. The following year, the company that had the mail contract for South Africa failed, and the Union line took over. The government contract specified that no ship under 530 tons should be used, so they began the service with just one vessel, the RMS *Dane*. She could scarcely be classed as an ocean liner, since she carried only three passengers on her first voyage, when she left England on 15 September 1857, arriving in Table Bay on 29 October. The contract had also specified that the journey should not take more than 42 days, so she only just made the deadline. It was a modest beginning for what would become a highly successful company.

In 1862, ship owner Donald Currie decided to start a service to India, but with sailing ships not steamers. They became known as 'Currie's Calcutta Castles'. By 1872, the route was being used by steamers carrying the mail and passengers and four years later it became the Castle Mail Packet Company. In 1876, the Cape government decided to use both the Union and the Castle companies, with the proviso that they should not combine. Their argument was that they would soon be competing to get the lion's share of the trade, which would ensure a good service to South Africa. It worked for a time and the two companies provided an extremely reliable service to South Africa. One outstanding vessel of the early years was the Union line's ship *Scot*, which on her maiden voyage in 1891 reached Southampton in 15½ days, with a stop at Madeira. In March 1893, she broke her own record with a time of 14 days, 18 hours, a record that remained unbroken for an astonishing 43 years. She was a beautiful ship, able to take 500 passengers with a service speed of 16.5 knots.

Once the terms of the mail agreement ran out, the two companies merged to form Union-Castle. Between them they established a regular service, designed to meet every traveller's needs. At precisely 2 o'clock in the afternoon, a ship would leave Southampton for Cape Town, and at exactly the same time a sister ship would leave Cape Town for Southampton. In 1912, Royal Mail line took over the company but kept the old name, and in 1922 they introduced the round Africa service, calling at twenty ports and taking nine weeks for the round trip. The routes were varied; if one voyage went down the west coast of Africa and returned through the Suez Canal, then the next would go the other way round.

A painting of the Union liner *Briton* by William Lloyd Wyllie.

In 1926, the Union-Castle ordered the first of a new line of 'Castle' ships with *Carnarvon Castle.* Steam was abandoned and replaced by two 8-cylinder diesel motors. She was the first of the line's vessels to exceed 20,000 tons. Perhaps to reassure passengers that the new ship was much like the familiar steamers, she was fitted with two funnels, but one was a dummy. As originally designed, she could carry a total of 850 passengers in three classes. She was given a refit in 1938, when the passenger list was reduced to 700 and the dummy funnel removed.

The 1920s saw organisational changes. Union-Castle bought up the White Star line, but with growing financial problems at the Royal Mail group, the end result was that Union-Castle became an independent company again and at once set about a five-year rebuilding programme. One of the new vessels, *Stirling Castle,* was powered by the largest pair of diesel engines ever built in Britain. The 10-cylinder engine stood 34ft high, was 72ft long and weighed 1,000 tons. They proved their worth when they powered the ship to a record-breaking run to Cape Bay of 13 days 9 hours, finally taking the honours from *Scot.* The building programme came to an end just as war broke out in 1939, and the regular services to South Africa only restarted in 1948. The company hoped that there would be a similar rush of immigrants to South Africa, just as there was to Australia and New Zealand. To meet that demand they built their first one-class ship, the *Bloemfontein Castle,* which set off

A postcard of the elegant Union-Castle liner *Stirling Castle,* built in 1935.

on her maiden voyage in 1950. Being one-class, there was no need for a whole series of separate public rooms varying in grandeur from first to utilitarian in third. Instead, there was a large lounge, reaching the whole width of the ship on the promenade deck, connected to a writing room and library and a smoking room. The 739 passengers could all be fed in just two sittings in the huge dining hall. But, sadly for the company, there was no rush of citizens heading for a new life in South Africa, and after just nine years in service, the ship was sold.

There were gradual improvements in the mail service, which by 1965 had reduced the passage time to just eleven days. More new ships had been added to the fleet, with steam turbines returning in place of the diesel engines. *Windsor Castle* was completed in 1960 as a two-class vessel, with each class having its own distinctive style of décor and furnishings. First class was described as 'traditional', which in practice turned out to be rather dull, but undeniably comfortable, with well stuffed armchairs in the public room. Second class was 'contemporary', with designs that seem to have been borrowed from the Festival of Britain of 1951. The circular smoking room, for example, had a motif on the floor that even seemed to echo the Festival official emblem. It is stylish, but the chairs look slightly less comfortable. But there was no shortage of amenities: air conditioning throughout; two swimming pools; a health spa; and a theatre that could also be used as a cinema. One other innovation came with the arrival of the *Transvaal Castle* in the 1960s, a one-class ship that employed female staff, somewhat coyly known as 'stewardettes'. But the days of the great ocean liners, offering a regular passenger service, were drawing to a close for both P & O and Union-Castle.

Chapter Ten

Safety at Sea

Throughout much of the period covered in this book, liners at sea had virtually no means of communication with the rest of the world once they were out of sight of land, unless they came near to another vessel. They could reasonably expect to avoid fixed hazards, provided they were marked on charts and navigation officers were doing their job efficiently. As we saw in the case of SS *Great Britain* that did not always happen, when the ship sailed blithely on past the Isle of Man and ran straight into the Irish coast. But, of course, navigation depended on being able to take measurements of the sun or the night sky, which with cloud cover might not be possible, and they could easily be carried off course. In those cases, one essential reading was the depth of water beneath the hull. This was tested using a length of rope, knotted at regular intervals, with a lead weight attached. The sailor would spin it round his head two or three times, rather like a cowboy with a lasso, then throw it overboard. He counted the number of knots passing through his fingers, until he felt the change when the lead hit the sea bed. This technique remained standard until the middle of the nineteenth century, when Lord Kelvin invented his 'sounding machine', a more sophisticated version of the old lead weight system, which in its more advanced stage of development included a tube with a chemical coating that changed colour with increased depth. A far more efficient idea was first developed by a German scientist, Alexander Behm, that he patented in 1913. It was the first echo-sounding device. It used a cartridge to sound an explosion and then measured the time taken for the echo to reach the surface, which as the speed of sound through water is a constant, gives the depth. Marconi introduced an improved version using electrically produced sound instead of an explosion. The first echo sounding device was used in the liner SS *Berkshire* of the German shipping line Militzer & Münch (M & M) in 1880. With good seamanship – and an element of luck – the ship would sail on without hitting any obstacles. Closer to shore, lighthouses provided warnings of dangerous rocks and reefs – and stone-built lighthouses had been built on many dangerous coasts from the middle of the eighteenth century.

If hitting solid objects could largely be avoided, colliding with other vessels proved more problematic. Looking through the history of the different vessels owned by the main companies, it is quite surprising to see just how many were in collision with other ships. In daylight in good weather, this was not a problem as lookouts were always posted and avoiding action could be taken. The rules of the sea are clear; both vessels should steer to starboard to pass port side by port side, and as a general rule, powered vessels give way to ships under sail. At night, the vessels would carry a bright white light, and two navigation lights: red on the port side, green to starboard. The greatest difficulty, however, came when the sea was shrouded with fog. On sailing ships, the best way of indicating presence was to ring a bell and blow a horn. On steamers, the steam whistle could blast out effectively. But it is surprisingly difficult to locate exactly where a sound is coming from in the fog and by the time the sound had been located, it could be too late to avoid a collision. So, collisions did occur, and the next objective was to try and minimise the risk of an accident turning into a catastrophe.

Wooden-hulled vessels were necessarily more fragile than iron-hulled, and where the two met it was the former that would always come off worst. In the case of the iron ship, however, a great deal could be done to limit damage. Brunel's *Great Eastern* was the first to be built with a double hull. This undoubtedly saved the ship from sinking in 1862 when she hit rocks off Montauk, Long Island, New York. They tore a 100ft long gash in the side of the hull, but only penetrated the outer layer; the inner hull remained intact and she was able to be rescued and repaired. There is a story, however, that has circulated for many years, that it was then that the repair gang discovered a skeleton inside the two hulls – a riveter who had been trapped there during construction. It was said that this made her an 'unlucky ship', which she certainly was in that she never made any money for her owners. It is highly unlikely that such an event ever took place. Riveting was a two-man job, and it is hard to imagine one man not noticing that his mate on the other side of the iron panel hadn't come out.

Over the years, the double hulls became more complex. The space between the hulls was divided into a series of watertight compartments. The *Normandie*, for example, had forty of these, while the *Queen Mary* had seventy. The double hull extended well above the water line, nearly forty feet in the case of the latter, creating an armoured honeycomb of protection. Further protection was supplied by watertight partitions across the full width of the ship, known as bulkheads. Access through the bulkheads was via watertight doors, specially designed to withstand considerable pressure – on the *Queen Mary* they were designed to withstand forces of up to thirty tons. It had largely been up to individual shipbuilders to decide just how many bulkheads were needed and how high above the water line they should reach. Following the famous sinking of the *Titanic* in 1912, of which more later, a special committee was appointed by the Board of Trade to set standards for British ships. The law then laid down regulations for the number required; for a vessel of 285ft length, four bulkheads were required and after that an extra one had to be added for every additional seventy feet.

In twentieth century ships, the arrangements for ensuring that the bulkheads did their job efficiently became an essential part of the safety programme. The heavy bulkhead doors could be opened and closed mechanically using hydraulics and a set of signals in the engine room and on the bridge showed which doors were open and which closed. In an emergency, all doors could be closed from the bridge in seconds, during which time a warning bell allowed time for those who needed to escape a section to get through. Although primarily intended to keep a vessel afloat if one section was holed, the bulkheads could also be closed to confine a fire.

Fire on board can have devastating results, so the more modern liners used a lot of fire-resistant materials throughout the ship. Some would certainly not be used today; all the cabins of the *Normandie* for example were lined with asbestos. Automatic fire detectors were installed in cabins and other enclosed spaces, with indicators on the bridge or in a special fire control centre to show exactly where the problem was situated.

The fire control room aboard the *Normandie*.

A major cause of disasters at sea was overloading, particularly in the nineteenth century in Britain. In just one year – 1871 – an amazing 856 vessels were lost off the British coast. Old ships were often insured for far more than their actual worth, so their loss was no financial hardship for the owners – it was indeed highly profitable. Instead of having a worn out ship to dispose of at the breakers' yard, they had the substantial insurance pay out. Sailors faced with what they called 'coffin ships' might refuse to go on board but would then face prosecution. There was an obvious remedy – the waterline mark on the hull; if that was not visible when the ship was afloat, then it was overloaded. It was not a new idea. In the late Middle Ages, the Venetian Republic, Genoa and the Hanseatic League of northern Europe had all introduced some sort of waterline mark on their ships. In 1835, Lloyds Register issued a recommended loading limit, but it was not legally enforceable. The increased loss of shipping led the MP Samuel Plimsoll to institute an enquiry into shipping losses. His aim was to make a waterline mark compulsory but faced strong opposition from many ship owners. However, the Merchant Shipping Act of 1872 was passed, but although all vessels now had to have a marking, its actual positioning on the ship was not legally established until 1894.

The waterline mark became known as the Plimsoll line in Britain, and it was modified over the years. Buoyancy depends on where the ship is trading; salt water for example, is denser than fresh water and cold water denser than warm. On modern vessels, the standard mark is a vertical line, with lettered horizontal lines for the different conditions with the least dense water at the top and the densest at the bottom. They are: TF for tropical fresh water; F -fresh water; T – tropical sea water; S- summer temperate seawater; W- winter temperate seawater; and WNA – winter North Atlantic. Markings are now required by international law, and this simple measure brought the age of the coffin ships to an end.

Perhaps the most important innovation not so much in avoiding accidents but in getting help when one occurred, was the invention of radio. Scientists, notably Clerk Maxwell and Heinrich Hertz, developed the idea of electromagnetic waves that would travel at the speed of light. In 1895, Ernest Rutherford succeeded in transmitting radio waves at Cambridge and receiving them three quarters of a mile away at a detector he had invented, but there was no immediate interest in the idea. The first really successful radio communication was the work of Guglielmo Marconi, who after initial experiments in Italy continued his work in Britain. He gradually increased the distances that signals could be transmitted. By 1895, he had received signals from a transmitter nine miles away and two years later set up

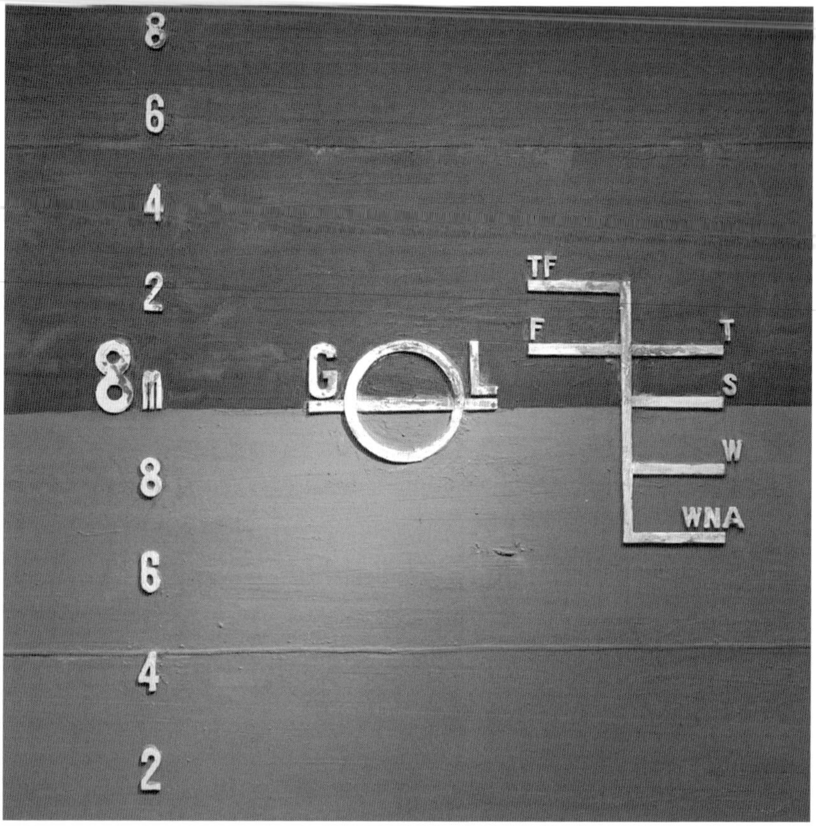

The Plimsoll line marking on a modern ship.

the Marconi Wireless Telegraph Company. The name is significant, for up to that time the only long distance form of communication that was comparable was the telegraph cable – this one, as the name told potential customers, needed no wires or any other form of physical connection. It was literally wireless. On 12 December 1901 he achieved his great triumph, sending a signal from Poldhu in Cornwall to St. John's, Newfoundland. The advantages of having such a system were immediately obvious and the value of radio communication was soon to have a dramatic demonstration. Even before Marconi's long distance demonstration, the Cunarder *Lucania* had been equipped with a radio in June 1900.

The liner *Republic* was built at Harland & Wolff and launched in 1903 and originally owned by a sister company of White Star, but was added to the main White Star fleet after just two voyages. She was fitted with radio. In 1904, she began operations between New York and Mediterranean ports. On 23 January 1909, she had left New York for Gibraltar with 742 passengers and crew on board. She hit a patch of thick fog off the coast of Massachusetts, and proceeded at normal

The radio operator on the *Oceanic* in 1913.

speed, sounding the fog whistle. Just before 6 in the morning, another whistle was heard nearby; the captain ordered the ship to go into full reverse and put the helm to hard-a-port. It proved to be too late, and the bows of SS *Florida* hit her midships. Passengers in the cabin where the bow hit died instantly, but it was at once clear that the damage had resulted in the engine room and boiler room flooding and the whole vessel began to list. The Captain put out a CQD call – the letters were later changed to the now familiar SOS. Some of the passengers were transferred to the damaged *Florida* and the American coastguard sent out a vessel as well. The *Republic* remained afloat for some time and later that day another vessel, the *Baltic,* reached the scene. Between them all the passengers and crew were saved – only six people died as a result of the collision. There were attempts to tow the *Republic* to safety, but she eventually sank on 24 January, the largest ship ever to have been lost at sea up till then.

One result of the loss of the *Republic* was the American Wireless Ship Act of 1910. This required all passenger ships carrying fifty or more passengers leaving an American port to have a radio capable

The damaged *Florida* after her collision with the *Republic.* The inset top left shows her restored after just 24 days in dry dock.

of transmitting at least 100 miles and to have a trained operator on board. The ability to send distress messages was only one advantage of having a wireless on board. General messages could also be sent and received. Ships could be warned of hazards, could receive weather reports and be kept in touch with the news. It can even produce some delightful surprises. My friend Nick Walker was the owner of an old coastal steamer, a Clyde Puffer that takes twelve passengers for trips around the west coast of Scotland. He changed the normal route to steam down to the Clyde to see the last visit of the *QE2* to the area where she was built. He was in sight of the great ship when he received a call from the captain of the liner, telling him how pleased he was to see an old Puffer still afloat. Nick was only able to say, in his surprise: 'A pleasure to see you too, sir'.

When things did go badly wrong and disaster struck, then everyone on board had to rely on some form of lifeboat or life-raft to escape. In the days of the sailing packets these were unlikely to be anything other than the ship's boats used for going ashore. There were no regulations covering provision of lifeboats, and a lack of them led to a tragedy. In March 1870, the paddle steamer *Normandy* was heading for the Channel Islands, with a crew of 28 and 36 passengers. She had a contract for carrying the mail, and was fined if late, so, although running into thick fog, continued on her course, sounding her foghorn. Then another warning sound was heard, but too late; she collided with the *Mary*, who ran into her starboard side. The captain ordered the two remaining boats on the port side to be launched and ensured that all the women on board were taken away, while he and the crew remained on board and died when the ship sank. At a subsequent enquiry, it was proposed that the government should make it compulsory for all vessels to have sufficient lifeboats to evacuate everyone on board. The government response was terse:

> In the opinion of the Board of Trade it will not be possible to compel the passenger steamers running between England and France to have boats sufficient to the numerous passengers they often carry. They would encumber the decks, and rather add to the danger than detract from it.

Consequently, no new legislation was considered and ships went to sea without any need to match lifeboat capacity to the numbers on board. By the end of the nineteenth century, however, rules were brought in, and one of the requirements was that ships with watertight bulkheads had to have lifeboats or rafts able to take half of those on board. This was one of the factors that contributed to the most famous accident ever to befall an ocean liner, the sinking of the *Titanic.* Her capacity was for a maximum of 2,224 passengers and crew, but the lifeboats only held about 1,400 at the most, which was actually better than the minimum legal requirement. Partly, this was because the *Titanic* was widely publicised as unsinkable thanks to the fifteen watertight bulkheads that reached to the top of five of the ten decks. It was assumed that in the case of an accident, at worst one or two compartments would be flooded and that she would still stay afloat. And even if there was a worse disaster, the boats would be sufficient to ferry the passengers and crew to nearby rescue ships.

Titanic set off on her maiden voyage on 10 April 1912, with a crew of 882 and 1,320 passengers; had she had her full capacity of 2,435, the tragedy would have been even greater. As it was, there was a feeling of complacency that nothing could go wrong on an unsinkable ship, and it seems that even the elementary precaution of having a boat drill so that everyone on board would know what to do in an emergency was ignored. Even more alarmingly, as the ship headed out into the

North Atlantic, warnings of icebergs drifting further south than was usual at that time of year were ignored. One of the great advantages of radio communication was precisely that ships encountering bad conditions could warn other vessels in the area. Three vessels reported the conditions, but word was never passed to the ship's officers on the bridge. Why was such important information withheld? In a curious arrangement, the radio operators on board were not employed directly by the line, but by the Marconi Company, and their first priority was to look after the passengers. There were many VIPs on board, anxious to send and receive messages, so the operators were kept busy. Were they too busy to actually note the iceberg warnings, or were they simply not aware of their significance? It has even been suggested that reception was bad, though that seems unlikely. Whatever the reason, *Titanic* sailed on at an almost full speed of 22 knots. When the lookout spotted an iceberg ahead at 23.39 on 14 April, the ship's bridge officers gave the order to go hard a-starboard and reverse the engines. The ship scraped along the side of the iceberg, and it hardly seemed to register on board that anything had happened other than a slight jolt. In fact, the collision had caused a rip in the hull, estimated to be at least 300ft long, opening up six of the watertight compartments. At once water came flooding in far faster than the pumps could remove it, and as the bows sank lower into the sea, so the water began flooding the decks

***Titanic* leaving** Southampton for what was to prove her last voyage.

above the bulkheads. The vessel was doomed. The captain gave orders to abandon ship, to muster the passengers to the lifeboats and for the radio operators to send out distress calls. Unfortunately, although the calls were sent out, they gave the wrong position for the vessel, out by 13 nautical miles. There was no public address system on board, so stewards were sent to rouse passengers and prepare to get them into the boats. The stewards in first class had only a few cabins each for which they were responsible and were able to get people out and help them to the boats. Those in the other classes had a much more difficult task, with far greater numbers to deal with. It is no surprise to find that the survival rate among the first class was far higher than for the rest. The rescue operation would have been farcical had it not been so tragic. As there had been no drills, no one seemed to know precisely what to do. The traditional order of 'women and children first' was interpreted differently by different boat crews, some interpreting it as not taking any men on board, so that as a result lifeboats left less than half full. Stories emerged later of the steerage passengers being locked away in the lower decks. Among those who did escape was the chairman of the White Star Line, Bruce Ismay. An hour after the ship went down, the *Carpathia* 'that had picked up the distress call' arrived on the scene to take on board the survivors. There were just 710 from the 2,202 on board. There was another significant statistic: 82 per cent of the first-class passengers survived, but only 26 per cent of third class.

Inevitably, there were enquiries into the disaster on both sides of the Atlantic. One result was the establishment of the International Convention for the Safety of Lives at Sea (SOLAS) in 1914, which recommended among other things that all ships needed to have enough lifeboats and rafts to carry all on board. It was never carried into effect, as a result of the outbreak of war the same year. A more comprehensive set of rules was established in 1929. In Britain, the Board of Trade produced new regulations in accordance with the SOLAS agreement. As well as insisting on a full set of lifeboats, it also ruled that every passenger should have a life-jacket that had to pass a specific test; it had to stay afloat for twenty-four hours while attached to a fifteen-pound lead weight. Although motorised lifeboats were already in use, officialdom was suspicious of such novelties. If there were more than fifteen lifeboats on a ship, one of them could be a motor boat and it had to be fitted with radio and a searchlight. The remainder were vessels that could be rowed and sailed. Things had moved on by the time the *Queen Mary* was launched, when all her lifeboats were motorised, each able to carry 145 people. There were rules too for the non-motor lifeboats. They had to carry two spare oars and a steering oar in case the rudder was damaged and a canvas sea anchor. They also

A *Titanic* lifeboat with passengers photographed by a passenger on the rescue ship *Carpathia*.

needed a compass, a supply of matches in a waterproof box, distress flares, an axe, lamps and lamp oil, buckets and a bailer. Provisions were basic, mainly biscuits and fresh water. Over the years, SOLAS has improved on the basic requirements of 1929. A modern lifeboat has to satisfy 31 requirements, including such useful items as sea-sickness pills and a fishing line. Life-rafts have 24 listed requirements.

During the Second World War, a new device became available for shipping – radar. In 1942 it became compulsory for all naval and large commercial vessels to be fitted with the device. Fog – the deadly source of so many fatal accidents – was no longer the impenetrable hazard it had once been. The name is an abbreviation of radio detection and ranging. Radio waves sent out from the ship bounce back from objects in their path and the results are shown on a screen giving the direction and distance of the object, whether it is a stationary obstacle such as a rock or iceberg or another vessel. Travelling by sea was at once made considerably safer.

Chapter Eleven

The Liners at War

When wars break out, one of the first logistical problems to be faced is getting troops to the war zone, and often this involved requisitioning civilian liners to do the job. The start of the Crimean War in 1854 presented the British with precisely that problem and was the first in which steamships were used as troop carriers, one of which was the SS *Great Britain*. By the end of the war, she had carried some 1,600 troops to the war zone. Horses were also carried and required special treatment. Just like human beings, they can experience sea sickness, but unlike us are unable to vomit, and this can lead to great distress and even death. To help with the situation, they were held in hammock-like slings that prevented them from swaying about too violently. They were also treated with vinegar, rubbed round their heads and nostrils which helped to alleviate the symptoms. They were fed and watered three times a day.

Brunel certainly would have approved the use of his great ship, for he was anxious to do everything he could to help the war effort. He proposed building a floating siege gun and sent his design to the Admiralty, who ignored it and even lost the proposal. He was, however, to make another contribution that probably saved more lives than any other measure taken during the war. Florence Nightingale had publicised the atrocious conditions at the military hospital at Scutari, conditions that led to the unnecessary loss of life of so many who went there for treatment. Brunel designed a prefabricated modern hospital in 1855 that was mainly built of wood, but the kitchen, bakehouse and laundry were iron to minimise fire risks. Two steamers, *Hawk* and *Gertrude,* were called into service to take the structure and the necessary tools, together with a dozen men and two foremen. They had expected help from the military in erecting the new hospital, but none was offered. The small team managed, however, to complete the work in just seven weeks. At the old Scutari site, half the soldiers who entered died. It was estimated that 1,500 were admitted to the Brunel hospital with just 40 deaths.

The main company involved in the war, however, was P & O. Twelve of their vessels were commandeered and between them they shipped over 60,000 troops and thousands of horses, together with supplies. Among the vessels was the *Himalaya,* built in Scotland in 1854, and the largest vessel in the fleet. She was so big, in fact, that after the war

The P & O liner *Himalaya*, one of the many vessels called in for duty as a troopship during the Crimean War, from a painting by J.W. Deering.

ended the company decided there would never be sufficient numbers of passengers to make her viable. They got rid of her, selling her to the government to begin a new life as HMS *Himalaya*. The same company was to be called on for an even greater effort shortly after the Crimean War had ended in 1856. The following year saw the start of the Indian Mutiny. Once again P & O stepped in as troop carriers and carried some 6,000 British troops. Another vessel heavily involved was the *Cape of Good Hope*, built in 1853 for the Southampton-Cape Town route. She was a screw steamer but also fitted out with full schooner rig. In 1856, she was bought by the Calcutta and Burmah Steam Navigation Company for their route between Rangoon and Calcutta. She had scarcely got started in her new civilian role before she was again requisitioned for troop transporting duties. She brought the first consignment of troops from what was then Ceylon (Sri Lanka) to Calcutta. Her working life was, however, short lived. In 1859 she was involved in a collision and sank. The Calcutta and Burmah later became the British Indian Steam Navigation Company, and with P & O were regularly called on for the different campaigns of Africa and Asia of the late nineteenth century, including the Afghan wars and the campaign in Abyssinia. The P & O fleet was showing its age by the end of the century and new vessels were brought into service, the *Britannia* and *City of Rome*. They offered considerably better accommodation than their predecessors and were notably faster – *Britannia* established a new record for the British Indian mail service, though, interestingly, while at sea she was once overtaken by the *Cutty Sark*, streaking through the waves with all sails set. But she too was to spend a good part of her career carrying troops to various campaigns at the end of the nineteenth century.

Early on in the twentieth century, nations began preparing for a major conflict. Germany was the first to think about how ocean liners might be used in wartime, and as early as 1908 orders were handed out to naval captains to prepare for converting them into auxiliary cruisers. One of the first to be armed was the *Kaiser Wilhelm der Grosse*, equipped with six 105mm and two 37mm guns. When Britain declared war in 1914 following the invasion of Belgium, she was sent to sea and succeeded in sinking three small ships, including a fishing vessel, *Tubal Cain*. All those on board were taken off first. Shortly afterwards, she encountered the Union-Castle passenger ship, the *Galician*. She was boarded by men from the German ship, but on discovering there were women and children on board, Captain Reymann signalled that he would not sink her. Her next encounter was with the Royal Mail Steam Packet Company ship, RMS *Arianza*. Once again, she was ordered to stop and was then asked how many women and children were on board, and on being told 335 women and 97 children, Reymann again allowed the ship to continue on its way.

The attack on the German liner *Kaiser Wilhelm der Grosse* in the First World War.

By now, the coal bunkers were in urgent need of refilling, so she pulled into the Rio de Oro where colliers were available. The process of refuelling was well under way when news came of the approach of the British cruiser, HMS *Highflyer*. Reymann set his prisoners on shore, and steamed out to sea, where at least he had a chance to manoeuvre, instead of being stuck in the river as a sitting target. It was an uneven contest; the German vessel was heavily outgunned and when she ran out of ammunition, the end was inevitable. There is some argument about what actually happened. The British Navy version is that their guns sank her; the Germans claimed that rather than surrender one of the greatest vessels ever to fly the German flag, she was scuttled on the captain's orders. Most of the crew were taken prisoner, but Reymann actually managed to swim ashore, and eventually made his way back to Germany. He is remembered mainly for his decency and sense of what was morally right, even during wartime.

The German liners were equally liable to be attacked by British ships. The *Kronprinzessin Cecilie* was en route to Bremerhaven with 1,000 passengers on board, when a curious message was received by the radio operator. 'Urgent and Confidential – Erhard has suffered attack of catarrh of the bladder. Siegfried.' The captain was not bemused; he had already received the instruction that this was a code to tell him that war had been declared. He at once ordered the ship to turn round and head back to neutral America to avoid capture or destruction by the British. Wealthy Americans on board offered him $1million if he would fly the American flag to ensure their safety, but he refused. He knew that the British had been told of his position by the French liner *La Savoie*. It was also well known that as well as passengers, the liner was carrying a fortune in bullion, and would be well worth pursuing. He set off at full speed for the nearest port, which turned out to be the fashionable seaside resort of Bar Harbor on Mount Desert Island, Maine. There she anchored and the captain and ship's officers, instead of becoming prisoners of war, were guests at the many parties being thrown by wealthy Americans. Not all liner and liner captains were that lucky.

Some British liners were taken over by the Royal Navy and were transformed into fighting machines. The Cunard *Carmania* was built in 1905 and represented the latest in marine technology, with steam turbines driving three propellers and with a capacity for 2,650 passengers. At the outbreak of war, she was transformed into a warship: the former Cunard colours vanished to be replaced by sombre black and grey, armour plating was added and the decks where the wealthy had once promenaded were now cleared and fitted with eight 120mm guns. She was sent to patrol the South American coast and headed for the uninhabited island of Trinidad off the Brazilian coast – no connection with Trinidad of the West Indies. There she found a German liner, the *Cap Trafalgar*, being bunkered. Like the *Carmania*, she was a modern triple screwed vessel, but with a smaller passenger complement of 1,600. She was, however, noted for the opulence of her interior design, but all that had now gone and she too had been transformed and fitted with two 100mm guns and six pom-poms. The German captain, realising his position and with no intention of being caught in harbour, at once set out to sea.

The Cunard liner *Carmania* fitted out as an armed merchant cruiser, illustrated by William Lionel Wyllie. The distinctive Cunard colours have been replaced by uniform grey.

The two ships manoeuvred for position and began to exchange fire. There were direct hits on both ships and the bridge of the *Carmania* was completely destroyed and numerous fires were started. It was to be the *Cap Trafalgar*, however, that was heavily outgunned and suffered the fatal blows and sank, but not before she was able to send out calls for help. The *Carmania* intercepted the calls and heard that the powerful battleship *Kronpriz Wilhelm* was on its way. The captain of the battleship, however, realised that when all signals from the *Cap Trafalgar* ceased that there was no ship left to rescue, and the *Carmania* escaped and was taken for repair. She would survive the war and eventually return to her old peaceful role as a passenger liner. No one could ever have imagined when those two splendid vessels were built that one day they would be pounding each other with shells and that one would be lost forever.

A new type of vessel was introduced into naval warfare in these years, the submarine. There had been developments of underwater craft as early as 1850, but with little success. It was only in the early years of the twentieth century that the modern version was built. The German Navy began commissioning their U-boats in 1906 and by 1912 they had introduced a very efficient version with diesel engine and two torpedo tubes. By the time the war started, they had 48 of these vessels in service. They provided a unique threat to all forms of shipping, including passenger liners.

In the early years of the war, there was a general agreement that passenger ships that were used by civilian travellers would not be attacked and that U-boat commanders would act as Reymann had done. This misconception led to one of the most famous sinkings of the First World War. The Cunard *Lusitania* continued to offer a passenger service across the Atlantic even after war was declared. She had an early escape on 4 August 1914, when she spotted an enemy warship but, as one of the fastest ships afloat, she was able to steam away and take cover in a bank of fog and eventually make her way safely to Liverpool. The situation at sea changed in November when the British government declared the North Sea a war zone, in an attempt to form a blockade, after which Germany retaliated by making a similar declaration about the seas round the British Isles. That was the position when the *Lusitania* was preparing to set sail from America on 1 May 1915. A clear warning was posted in the press, warning potential passengers of the risk they were running, explaining they would be entering the war zone; 'vessels flying the flag of Great Britain, or of any of her allies, are liable to destruction in those waters and that travellers sailing in the war zone in ships of Great Britain and her allies do so at their own risk'. It could hardly have been made clearer. Yet an old newsreel shows crowds of well-dressed passengers arriving in fleets of taxis ready to board the ship, and one of them remarked cheerfully to the cameraman that he'd be watching the film when he got to London. As the ship left the port, the decks were crowded and one passenger waved the stars and stripes, for many of the 1,266 passengers on board were Americans. On the evening of 6 May, the ship received an ominous message that there were U-boats off Fastnet, and advised the captain to keep to the middle channel and proceed at full speed. On the morning of 6 May she was 120 miles off the Irish coast, when she entered fog and was forced to reduce speed and sound her fog horn. By noon the fog had lifted, and she was able to return to full speed, but now she had been spotted by the submarine U-20 that submerged and set a plot to intercept her. An American businessman, James Brooks from Connecticut, gave his own account of what happened at 2.40 that afternoon. It is quoted in Brinin's book, *The Sway of the Grand Saloon*.

"All the News That's Fit to Print."

The New York Times.

EXTRA 5:30 A. M.

VOL. LXIV...NO. 20,923. NEW YORK, SATURDAY, MAY 8, 1915.—TWENTY-FOUR PAGES. ONE CENT

LUSITANIA SUNK BY A SUBMARINE, PROBABLY 1,260 DEAD; TWICE TORPEDOED OFF IRISH COAST; SINKS IN 15 MINUTES; CAPT. TURNER SAVED, FROHMAN AND VANDERBILT MISSING; WASHINGTON BELIEVES THAT A GRAVE CRISIS IS AT HAND

SHOCKS THE PRESIDENT

Washington Deeply Stirred by the Loss of American Lives.

BULLETINS AT WHITE HOUSE

Wilson Reads Them Closely, but Is Silent on the Nation's Course.

HINTS OF CONGRESS CALL

Loss of Lusitania Recalls Firm Tone of Our First Warning to Germany.

CAPITAL FULL OF RUMORS

Reports That Liner Was to be Sunk Were Heard Before Actual News Came.

Special to The New York Times.

WASHINGTON, May 7. — Never since that April day, three years ago, when word came that the Titanic had gone down, has Washington been so stirred as it is tonight over the sinking [illegible] ports told that there had been no loss of life, but the relief that these advices caused gave way to the greatest concern late this evening when it became known that there had been many

The Lost Cunard Steamship Lusitania
X Where the First Torpedo Struck. XX Where the Second Torpedo Struck

SOME DEAD TAKEN ASHORE

Several Hundred Survivors at Queenstown and Kinsale.

STEWARD TELLS OF DISASTER

One Torpedo Crashes Into the Doomed Liner's Bow, Another Into the Engine Room.

SHIP LISTS OVER TO PORT

Makes It Impossible to Lower Many Boats, So Hundreds Must Have Gone Down.

ATTACKED IN BROAD DAY

Passengers at Luncheon—Warning Had Been Given by Germans Before the Ship Left New York.

Only 650 Were Saved, Few Cabin Passengers

QUEENSTOWN, [illegible] day, May 8, 4:28 A. M.—Survivors of the Lusitania who have arrived here estimate that only about 650 of those aboard the steamer

The sinking of the *Lusitania* made headline news.

I had just finished a run on deck, when I glanced over the water. It was perfectly smooth. My eyes alighted on a white streak making its way with lightning like rapidity towards the ship. I was so high in that position above the surface of the water that I could make out the outline of a torpedo. It appeared to be about twelve feet, and came along possibly three feet below the surface, its sides white with bubbles of foam. I watched it passage, fascinated, until it passed out of sight behind the bridge, and in another moment came the explosion. The ship, recoiling under the force of the blow, was jarred and lifted, as if it had struck an immovable object. A column of water shot up to the bridge deck, carrying with it a lot of debris, and despite the fact that I must have been twenty yards from the spot where the torpedo struck, I was knocked off my feet. Before I could recover myself, the iron forepart of the ship was enveloped in a blinding cloud of steam.

Shortly afterwards, the ship began to list to starboard, which made it impossible to reach the lifeboats overhanging the water, far out from the side of the ship. Portside lifeboats were launched, but they scraped against the side of the vessel, damaging them even before they reached the sea. Of the forty-eight lifeboats on board, only six were successfully launched. Others that were launched out tipped over due to the crush of people trying to get into them. The surviving lifeboats managed to pick up a few swimmers. Among those who were saved was the captain who was swept off the bridge by a great wave, but managed to find a floating chair to cling onto when he hit the water. He was fortunate. Of the 1,959 on board only 764 were rescued. One hundred and twenty-eight Americans were among the dead, which produced an angry response from Walter Hines, the American Ambassador in London: 'The United States must declare war or forfeit European success.' They did not, until April 1917, but the sinking of the *Lusitania* did a great deal to swing public opinion there against Germany. It has to be said that the German navy had only done exactly what it had said it would do – treat any British ship as hostile and attempt to sink it. It was also revealed that the ship had been carrying some military material, including ammunition. There was, however, no excuse for another sinking of a former liner.

The Union-Castle ship was originally named *Galician* when she was launched in 1900. In 1914, she was requisitioned by the Royal Navy as a hospital ship and renamed *Glenart Castle.* Like all hospital ships, she carried clear markings, including a large red cross painted on the side. On 25 February 1918, she left South Wales, heading for Brest. Shortly afterwards she was hit by a U-boat torpedo, which not only fatally damaged the ship, but destroyed most of the lifeboats. She sank in less than ten minutes and the crew only managed to launch one lifeboat and as a result, only 29 of the 182 on board survived and were picked up by a French vessel. There was outrage at the sinking, which was against international law and contrary to the orders given to U-boat captains. But worse was to follow. Another Union-Castle liner, *Llandovery Castle,* was also converted into a hospital ship. On 27 June 1918 she was torpedoed by U-86. The captain of the U-boat seemed to be aware that he had just committed a crime and decided that the best solution was to make sure that no one survived to give evidence against him. He rammed one lifeboat that had several nurses on board and machine-gunned many struggling in the water. Only one lifeboat got clear with just 24 on board. After the war, there was a demand that the U-boat officers should be charged with a war crime. Two officers of the U-86 were brought to trial and sent to prison for four years but were later pardoned on the grounds that they had only been doing what their captain ordered. The captain himself, Helmut Brümmer-Patzig, escaped to what was then the independent city of Danzig, where he was able to remain and avoid prosecution.

It was not only British ships that were subjected to U-boat attacks. The German steamer of the Hamburg-American line, *Abraham Lincoln,* was in New York when America entered the war. The German crew tried to cause as much damage to the ship as possible to prevent her reuse by the Americans, but she was repaired and went into service as a troop carrier taking soldiers to France. She made a total of five trips to France carrying over 20,000 troops. On her return from the last voyage, she and other troop ships were escorted by destroyers until they were considered to be outside the area where the U-boats operated. That turned out not to be the case. On the morning of 31 May 1918 she was struck by three torpedoes and sank within 20 minutes. Fortunately, almost all of the 715 on board made it into the lifeboats, where they were picked up by destroyers: Twenty-six men died in the initial attack and one officer was picked up by the U-boat and taken prisoner.

Some fifty large passenger steamers were lost during the war, mainly by U-boats, but a few hit mines. Among the latter was the former White Star liner *Laurentic,* that had been converted into an armed merchant cruiser, serving mainly in the Far East, but in January 1917 she was back in British waters and set sail with a secret cargo of bullion that was intended to be used to buy armaments from neutral America. On 17 January she struck a mine off the coast of Ireland and sank. The crew were able to take to the lifeboats, but the weather was atrocious, a blizzard was blowing and the temperature dropped to -13°C. Tragically, having survived the explosion, 354 of the crew died from hypothermia.

One of the many ships that sank deserves a special mention. The *Carpathia* is best known for her role in coming to the assistance of the *Titanic*. Like other liners, she was pressed into war service as a troop carrier. On 15 July 1918 she left Liverpool for Boston to collect troops, with a crew of 166 and just 57 passengers. She was part of a convoy, escorted for two days, but as the escorts left, the convoy was split in two, with *Carpathia* leading one half. It was then that she was hit by three torpedoes and began sinking. All but five managed to escape but were in danger of being attacked by the U-boat. Fortunately, a British warship arrived on the scene, and all were saved.

When America entered the war, one of the first actions was to commandeer the German liner *Vaterland* in 1917 to convert her into a troopship and rename her *Leviathan*. She made her first voyage under her new name in December 1917 and made 19 crossings, carrying around 100,000 American troops. They would not ever perhaps have thought that the chief quartermaster would end up in Hollywood, but he did. His name was Humphrey Bogart. After the war, she remained an American ship, returned to normal passenger duties.

Throughout the war, the liners did hugely valuable work but too many failed to survive to enjoy the peace. Those that did were soon

The liner *Mitava* as a hospital ship in 1918.

back to their more familiar role as luxury liners, and enjoyed what many regard as the golden age of the ocean liner, that lasted until the threat of war reappeared in the late 1930s. Preparations were made for transferring liners to war duty just as they had been in 1914. In Britain, protection against U-boat attacks would be provided as far as possible by sending ships out in convoys, with warships as escorts. Liners were to be 'disguised' using a technique that had first been developed in the First World War. This was known as dazzle camouflage, painting the hulls with zebra-like patterns in the hope that the ship's outline would be more difficult to identify from a distance. The one exception was the *Queen Mary*. Because of her superior speed, she made Atlantic crossings unaccompanied, painted a dull grey rather than covered in dazzling stripes. She had a highly successful career as a troopship, sometimes packing in as many as 10,000 soldiers on a single voyage and carrying some 250,000 over the course of the war years.

Inevitably, there were casualties and one of the more controversial occurred right at the start of the war. The SS *Athenia* of the Donaldson Atlantic line set off from Liverpool for Canada on 1 September 1939 with 1,103 passengers, including around 500 Jewish refugees and 311 Americans, who at this stage were not at war. On the evening of the 3rd, the day war was declared, she was spotted by U-30 that tracked the liner for a while before firing two torpedoes, one of which hit the port side by the engine room. Altogether, 98 passengers and 19 crew

The *Mauretania*, given dazzle camouflage in the First World War.

died in the attack. Tragically, some of the deaths occurred during the rescue operation. Lifeboats had been lowered and several ships responded to the distress call. One of the vessels was a Norwegian tanker, and one of the lifeboats fell foul of her propeller while another capsized in the heavy swell. The remainder of the passengers were brought safely to Ireland. American citizens were among the dead, and Germany feared a repetition of the furore following the sinking of the *Lusitania*. They issued a statement that there was no U-boat anywhere near the incident, but as it was impossible to imagine who else would torpedo a vessel in these waters, the story was not widely believed. Stories did, however, circulate that she had actually hit a British mine and the story of German involvement had been made up to persuade America to join the war. The German involvement was not officially confirmed until 1946.

There was a plan in Britain to remove children from areas of Britain that were being heavily bombed on a regular basis. Some were to be evacuated under the Children's Overseas Reception Board (CORB) scheme, which sent them overseas. The *City of Benares* was a steam turbine liner built for the Ellerman Line, that in peacetime ran between Britain and India. Now she was given a new role, mostly carrying refugees to Canada. She set sail on 13 September 1940 with 406 on board including 90 CORB children and their 10 adult escorts. Although she was part of a convoy, she was one of the lead ships and was hit by two torpedoes on the 17th and began sinking. There were difficulties in lowering the lifeboats and only 148 survived including only 13 of the children being sent, it had been hoped, to safety. But it was not only children of Britain and her allies who suffered similar tragedies. The Japanese passenger ship *Tsushima Maru* was taking civilians being evacuated from Okinawa to Kagoshima on 24 August 1944 as part of a convoy that was attacked by the USS submarine *Bowfin*. She sank with huge loss of life, including 775 schoolchildren – only 59 children survived.

A survivor of the *Athenia* being helped aboard the Norwegian tanker *Knute Nelson*. (Press Agency)

A number of German liners were, like the British, converted for military use. Some of their greatest losses occurred during Operation Hannibal that began in January 1945. By this time, it was clear that Germany was losing the war and Admiral Dönitz ordered a mass evacuation of German troops from Poland and the Baltic and a vast fleet of military and former merchant ships was assembled, among them the 1937 cruise liner *Wilhelm Gustloff*. She left Gotenhafen (now Gdynia) on Poland's Baltic coast on 30 January. No one has ever been sure just how many were on board; the passenger list cited 6,050 but more joined unrecorded – final estimates are that when she set out there were over 10,000 passengers and crew on board. There were wounded soldiers and civilians on board. Soon after leaving port, she was attacked by the Russian submarine *S-13* that fired three torpedoes. Only nine lifeboats could be lowered – most were frozen into their davits. Some tried to escape by leaping into the freezing cold sea, where their chances of survival were virtually nil. It has been estimated that over 9,000 died that day. It remains the largest loss of life from the sinking of just one vessel in the whole of maritime history.

At the end of the Second World War, the liners that survived returned to normal duties after suitable overhauls to make them fit for civilian service. Everyone thought that these splendid vessels would never have to be used in wartime again. They were wrong. On 2 April 1982 Argentinian forces invaded the Falkland Islands, or as they are known in South America, the Malvinas. As the total population of the island was only around 3,500 and had no army, they had no problem in taking possession. At the time, few people in Britain probably knew that the Falklands were British or even where they were, but nevertheless a war was declared and a task force sent out to regain the lost territories. The *QE2* was commissioned as a troop ship. It was a complex operation, as the ship was not designed to have enough fuel for an 8,000 mile journey. She had to have special arrangements for oil to be piped on board from a tanker at sea. She took on more fuel and water at Freetown, Sierra Leone. There was one further stop at Ascension Island before she reached South Georgia. Here the troops were transferred to

A gun mounted on the German liner *Bremen*.

***Canberra* with** her escort vessel HMS *Andromeda* near Port Stanley in the Falklands, shortly after the Argentinian surrender.

the *Canberra*. So far, it had been a remarkably comfortable voyage for all on board, enjoying the luxury of a great liner. Now, however, there was a grim reminder of what lay ahead, as wounded from the attacks on the naval ships *Coventry, Antelope* and *Ardent* were brought aboard for the journey back to Southampton.

The *Canberra* had already made one voyage to the Falklands with 4,000 troops, half of whom were transferred to smaller craft for the actual landing, not as the Argentinians had expected at Port Stanley, but on the eastern coast of East Falkland. It was here that the fleet was attacked by bombers and missiles causing the casualties that were then taken to the *QE2*. They had to return to the war zone, but once again escaped without damage. By 14 June the British had regained the Falklands – the war was over.

Chapter Twelve

Competition

At the end of the Second World War, European fleets had been greatly depleted, and many lines had to rely on their older vessels to maintain services. As the economies of all had taken a hammering in the war – and that included Britain, which had huge war debts – there was no question of building new vessels for some years. America, however, was in a far better position. The country had been on the winning side and had not been occupied nor had her mainland territory attacked. Her industries remained intact, and she was able to spend a fortune on a new and magnificent vessel, the SS *United States*. She was built between 1950 and 1951 at a cost of $79.4million – around $500 million at today's prices. The government put up $50million but with the proviso that the design would allow her to be easily converted into a troop carrier or hospital ship if necessary. Fortunately, that need never arose.

Her interior design was discussed on page 68, but that was not her only new feature. Her construction was quite different from that of earlier vessels, relying on prefabrication, with parts being fabricated and brought to the dry dock in Newport News, Virginia, for assembly – a grand total of 183,000 items. Her steam turbines, the most powerful built at that time, drove four propellers, giving her a cruising speed of

The *United* *States* leaving Britain for her return to New York on her maiden voyage.

35 knots and a range of over 10,000 miles. Not surprisingly, she broke the Atlantic speed record on her maiden voyage, when she averaged 35½ knots for the outward voyage and 34½ for the return; she took the blue riband. With a journey time of just three and a half days and a luxurious, modern styled interior, she was a popular ship. She was able to carry almost 2,000 passengers, but she is mainly remembered as the vessel that brought the stars of Hollywood and Broadway to Europe, such as Marilyn Monroe and Cary Grant. Walt Disney joined the ship and Duke Ellington travelled with his band. Among the staff when Ellington came on board was a steward, Claude Jones. He too had been a highly successful jazz musician, a trombonist who had played with the best in the business from Cab Calloway and Louis Armstrong to Coleman Hawkins and Zutty Singleton and had been in the Ellington band twice, from 1944-48 and again in 1951. It would be good to think he had a chance to change out of his uniform for another blow with Ellington while the band was on board.

Cunard were still relying on the *Queen Elizabeth* and *Queen Mary* to keep their transatlantic services running until they could afford to build a new vessel – and when the material was available with which to build it. The ship was finally launched in 1967 from the John Brown shipyard on the Clyde and was named by the Queen as *Queen Elizabeth 2*. One can see why the name appealed, but it caused a problem for Cunard, especially for a ship built in Scotland – for there had never been a Queen Elizabeth 1 on the Scottish throne – the Tudor monarch only reigned over England and Wales. The solution was to shorten her name to *QE2*.

She was originally powered by steam turbines, which gave a certain amount of trouble and were replaced by two diesel electric units to drive her twin propellers. Unlike the earlier Cunard liners, the steel hull was welded, not riveted. She was to have many alterations over the years that we shall come back to later.

The French were in a similar position to the British, in that their fleet now consisted of the ageing *Ile de France* and *Liberté*. Originally it was thought that the solution would be to build two vessels for the Atlantic trade, each of around 35,000 tons, but eventually it was agreed instead to build one splendid vessel of twice that size that would do for French prestige what the *Normandie* had done for a previous generation. That ship was the SS *France*, at over 1,000ft long overall, the longest liner built at that time. This was no accident, but a deliberate design decision – if she was to do France proud, she had to be the biggest. She was a beautiful ship, distinguished by her two funnels with wings to deflate the exhaust fumes from the deck areas. The previous *France* of 1912 had been famous for its opulent, overblown décor, but since

The *QE2* at Trondheim.

then the *Normandie* had revolutionised design ideas with its art deco interior. Now new ideas were needed to fit the second half of the century. In the event it was not so very different from the *United States* and the *QE2*, with much use of plastics and aluminium. But, being a French vessel, one area was given very special attention. No ship ever possessed a grander galley, and none produced finer food. One critic declared that the Grille Room was as good as any other restaurant in France, which to the French, of course, also meant best in the world. She was designed for just the two classes, first and tourist. The one place where they might have mingled was the theatre, but this was the one double-height space on the ship, and the only entrance to the balcony was on the first-class deck and tourist class could only use the stalls. Like the *QE2,* she was to have a varied career during her working life.

The German Norddeutscher Lloyd line was in a far worse position than the other lines described; their headquarters had been bombed

The *France* at Le Havre in 1978.

and their fleet was no more. They returned to the Atlantic but had to make do with second-hand ships, starting with the purchase of the Swedish vessel *Gripsholm*, which when she was built in 1924 was the first ocean liner to be powered by a diesel motor. She was renamed *Berlin* and began her service on the North Atlantic run in 1955.

The routes to the Far East run by P & O did not suffer quite such a heavy rate of loss as some other lines, though they were bad enough, but as a result they still had a number of vessels available to continue in service after the war. There was no shortage of passengers, particularly emigrants headed for Australia. Some of the ships, however, were decidedly close to their sell-by date. The *Ormond* was actually launched in 1917 and saw service in both world wars. In the first she served as a troop carrier, and then went into service to Australia with the usual mixture of first and second-class passengers, but in 1933 she was converted to just the single tourist class. In 1939 she was back as a troopship again, first taking part in the troop evacuations from France and Norway, then in the landings of North Africa, Sicily and Italy, finishing the war based in Bombay. In 1947 she was sent back to Cammel Laird at Birkenhead to be fitted out as an immigrant ship, with her accommodation increased from 777 to 1,070 third class passengers.

Two identical liners, *Mooltan* and *Maloja,* were completed in 1923 and both saw service in the war. They were initially fitted out as merchant cruisers but were later used as troopships. *Mooltan* almost failed to survive, having been bombed by German aircraft on her way to Freetown, but only suffered minor damage. Both were returned to their civilian roles as tourist class liners. Other venerable craft that survived to re-enter service were the *Orontes* of 1929, *Carthage, Strathmore* and *Orion* of 1935 and *Orcades* of 1937. The grandest of them all was the *Strathnaver,* built in 1931, one of the fleet known as the White Sisters from their gleaming white hull and superstructure. She was turbo-electric and had a very distinctive outline, with three rather stumpy buff-coloured funnels, though in fact only the middle funnel served a useful purpose. She was returned to P & O and adapted to carry 1,252 passengers, virtually all immigrants. During the refurbishment process, the two dummy funnels were removed.

P & O began building up their fleet after the war, starting with the *Orcades* of 1948, designed for the route to Australia, at first through the Suez Canal, but later changing to a new route through the Panama Canal, also serving ports on America's west coast. The next addition to the fleet was the second of the line to be named *Chusan*. She almost got off to a disastrous start at the launch in 1949, when the cradle supporting the hull broke, but all went well, and by the end of the year, she had made her first voyage to India, a route that was extended to Japan in 1952. She carried 474 first class and 514 tourist class, all of whom benefitted from an innovation. The Denny-Brown stabiliser had first been tested by the Admiralty in 1939 and installed in warships. This was its first application to a liner, and greatly reduced the effects of heavy seas.

One ship was to become famous for just a single voyage, otherwise she had very little to distinguish her from many others. She started out as the German motor vessel *Monte Rosa* in 1931, as a passenger

The *Chusan* – the second P & O liner to bear that name.

liner for the Hamburg Süd line with her first voyage to South America. Like many other vessels, she was converted to a troop ship in the Second World War. After the war, she was taken over by the British Government and renamed *Empire Windrush*. She eventually finished up being run by the New Zealand Shipping Company and was used on routes between Britain and the Far East and eventually to Australia. It was on a return trip from Australia in 1948 that she called in at Kingston, Jamaica. She was still owned by the government, and she was there to pick up servicemen on leave. Jamaica was still part of the British Empire in those days and so the population was officially British and allowed to come to Britain to live. A few locals decided to book a passage, but the ship that was designed to carry over 1,500 passengers was almost empty. An advert was put in the Kingston paper, *The Daily Gleaner*, offering passage to Britain at bargain prices. For the luxury of a cabin the cost was just £48, but if you were prepared to rough it in the accommodation supplied for troops in its war service years, it was a mere £28. For many Jamaicans this was the ideal opportunity to leave the poverty of the Islands for what they hoped would be a better and more prosperous future in Britain. Exact figures are not known but over 800 on board were registered as coming from the Caribbean to which number a few stowaways could be added. Those who made the voyage became known as the Windrush Generation. They did not receive the welcome they had hoped for – and deserved – but that is another story.

This was the only voyage involving immigrants, but the vessel continued in service for a few more years. Then, in February 1954, she set off from Yokohama but was plagued by problems – so that it took her ten weeks to get as far as Port Said. On 28 March, in the Mediterranean, there was an explosion in the engine room and all electrical power was lost. The emergency generator provided a little power, but not enough to quench the fire. Life boats were lowered, where possible – though some could not be lowered. Miraculously, all but four of the 222 crew and 1,276 passengers survived, and the unfortunate ones had died in the explosion. Attempts were made to salvage her, but she sank on 30 March. It had been a memorable career for the ship, that had been everything from a luxury liner to a troopship, but it is that one voyage of 1948 that earned her a special place in the history of Britain and the West Indies.

The one decisive factor in the continuing use of liners on scheduled services was that for the first half of the twentieth century there were no viable alternatives for ocean travels. There had been a threat to the lucrative first-class passenger trade early in the century. Lighter than air craft had appeared in the shape of balloons as early as the eighteenth century, but there was no means of steering them – they went where the wind took them. In order to make them useful forms of transport, they needed to have a source of power, and the nineteenth century saw experiments, first with small steam engines and later with electric motors. It was only with the arrival of the petrol engine, however, that the powered airship became a practical form of transport. The name forever associated with later developments is that of Ferdinand Graf von Zeppelin. His airships became all too well known in Britain, during the First World War, when they appeared over Eastern England on bombing raids. After the war, Zeppelin turned to civilian transport and they began building far larger craft – airships to cross the Atlantic.

The most famous and most successful Zeppelin airship was the *Graf Zeppelin*. This immense craft was 237 metres long, but the gondola slung underneath the main frame was just 30 metres long and could only accommodate 24 passengers. They enjoyed something of the luxury of the liners, with similar art deco design and the real appeal was the lounge and bar area with its windows, from which, no doubt with a feeling of superiority, they could look down on the ships passing below them as they zoomed overhead at almost 80 miles an hour. The airships were never a serious threat to the liners, simply because of the small numbers they could carry and the expensive fares. That threat

The *Graf Zeppelin* being welcomed back to Germany after a round the world flight. The airship was the first alternative option for passengers crossing the Atlantic.

disappeared altogether when in May 1937 the Zeppelin company's biggest ever craft, the *Hindenburg,* crashed in flames. It was not only a disaster that made the headlines, but it was seen on film by millions. Confidence in airships evaporated. That threat was gone but as the airship was being developed, so too was the heavier than air craft, the aeroplane.

There was no real attempt to begin passenger services across the Atlantic until 1936, when both British Imperial Airways and Pan American began services using flying boats. The most impressive was the Boeing B-314, a flying boat able to carry 74 passengers. It went into service with Pan Am in 1939 but had hardly had time to establish a regular service before war brought everything to a halt. After the war, the Americans resumed services, this time using conventional aircraft rather than sea planes, at the end of 1945. The flights by DC4 planes between New York and Hurn Airport near Bournemouth were expensive – and required two stops, at Newfoundland and Ireland. They were not seen as a serious threat to the established ocean liners at this time. A similar story can be told of the flights to the Far East by Imperial and Pan Am using flying boats. By 1939, Imperial were offering three flights a week, using Short flying boats from Southampton to Australia. It was not a notably speedy trip, with ten days in the air and nine stops along the way. It must have been quite an enjoyable experience, as the planes flew comparatively low and slowly, so there was always plenty to look at along the way, and sightseeing at the stops. Once again, war ended the services. After the war, Imperial ceased to exist and the routes were taken over by the state-run BOAC. It was BOAC that changed everything.

In 1952 BOAC introduced the world's first passenger jet, the de Havilland Comet. The pressurised cabin made for an experience very different from that of piston-engined aircraft – quieter and no vibration – and was fitted with what were known as 'slumber seats' that reclined. Most importantly, journey times were slashed. When introduced on the route from London to Tokyo, the flight time was 36 hours, compared to the previous best of 86½ hours. It was a new age and seemed to offer British manufacturers a world lead. Then came a series of accidents, that were later found to be down to metal fatigue. The lead was lost, but an example had been set, and the new age of the jet airliner was well and truly established and grew dramatically. In 1976 Boeing introduced their 747, the jumbo jet, that could carry up to 524 passengers in two classes. No need now for stops en route when crossing the Atlantic – the days taken by the ocean liners had shrunk to hours in the air. It was a competition that ultimately they would be unable to beat.

Chapter Thirteen

The Liner in Fiction

There are few novels written about the early days of passenger travel in the age of sail. William Golding, however, not only managed to write a novel almost entirely based on board a ship leaving England for Australia with cargo, a few passengers and emigrants, but extend the story over three volumes, beginning in 1980 with *Rites of Passage* and followed by *Close Quarters* and *Fire Down Below*. No date is given in the book, but from the description it is obviously set in the early nineteenth century. They are generally accepted as modern masterpieces but obviously a successful series of novels must have more to them than a bald account of a voyage – there has to be drama, and they have that in plenty. They do, however, tell us a great deal about the perils of voyages in those days and how the sailors dealt with them, for example, when they are caught in a storm with towering waves or having sailed too far south are threatened with destruction by icebergs. It is to be hoped that not every passage to Australia was as full of drama as this one. If this had been an account by a contemporary, anyone reading it who had been considering taking a ship to Australia would have started making new plans.

One other novel written at a similar time, Patrick White's *A Fringe of Leaves* (1976), also deals with a ship travelling between Britain and Australia, but this time in the opposite direction. Set in the 1840s, the voyage itself only takes up a small part of the story, but the crucial event that drives the drama is a shipwreck and the story of two of the survivors. Shipwrecks and disasters at sea seem to be very attractive subjects for several writers of fiction.

The arrival of steam powered liners has proved a far more popular source of inspiration for writers. The liner is an enclosed world for the length of time it is at sea – time for all kinds of interactions between those aboard. One of the earliest is Jules Verne's *Une Ville Flottante*, 1874 – published in English as *The Floating City*. Set around a description of a passage on Brunel's *Great Eastern*, between Britain and America, it has been quoted earlier (p74) for its description of life aboard the famous ship. Being a novel, however, not a travel book, there has to be something more dramatic than just an everyday account of how the passengers spent their time. This time the drama comes to a head in a somewhat unlikely event,

The cover design for Jules Verne's novel based on a voyage on the *Great Eastern*.

when the captain agrees to two of the passengers fighting a duel on the deck – a duel with an even less likely outcome. It has to be said, this is not Verne's greatest work, but it is a fascinating account of the *Great Eastern* itself.

Returning to the theme of shipwrecks, if the modern reader starts a novel featuring a huge liner, built out of steel with watertight compartments that make it 'unsinkable' – and then you add in the name of the vessel, *Titan* – then that reader will expect it to hit an iceberg somewhere along the way. And sure enough, that is exactly what happens and the unsinkable sinks. But there is a surprise here; the novel *Futility* by Morgan Robertson was published in 1898 before the *Titanic* was even built. Following the famous disaster, the novel was republished and somewhat cynically given a fresh title – *The Wreck of the Titan*. There is probably scope here for some conspiracy theorist to come up with something that makes all this less than a mere coincidence.

The *Titanic* itself has always attracted the film industry, and the first movie appeared just a month after the disaster. *Saved from the Titanic* was actually written by a survivor, Dorothy Gibson, who also starred in it. As far as is known, no copies have survived. 1912 also saw the arrival of a German silent movie *In Nacht und Eis*, also known as *The Sinking of the Titanic*. The first sound film about the disaster was originally called *Atlantic* as the White Star line prevented the use of the name of the actual ship. It was made in three languages – English, French and German – and although White Star objected to the name, filming did take place on another of their liners, RMS *Majestic*. It was, in terms of box office, a great success, but considerably less successful as an accurate portrayal of what actually happened in that famous disaster. Quite the oddest film was a German version made in 1943 as Nazi propaganda. All the British officers on board behave in a cowardly fashion and only the German first officer acts heroically to save passengers. There was, of course, no such officer on board. 1953 saw an American version of the story, starring Clifton Webb and Barbara Stanwick, but by far the best film to appear up to then was the British film of 1958, *A Night to Remember*, based on the book of the same name by Walter Lord. The book had been carefully researched, with Lord interviewing many survivors and the director Roy Ward Baker stuck closely to the factual accounts. Of the 1980 film *Raise the Titanic* there is little to say, except that it was a disaster movie in the wrong sense of the term – a total box office flop. After that, we finally arrive at the *Titanic* film everyone knows, James Cameron's epic of 1997, hugely expensive and equally hugely successful, collecting eleven Academy Awards. One might think it was the last of the long succession of movies on the subject, and most of us know nothing about anything that came after. There have, however, been three animated films, the first of which '*The Legend of the Titanic*' was made by an unlikely co-production team of North Korea, America, Spain and Italy in 1999. The sequel was purely North Korean – and in between those two came a Spanish-Italian co-production. One book does deserve a special mention: David Dyer's *The Midnight Watch* (2016) takes a very different approach to the story. The watch in question is the officer on board the *Californian* who saw the distress signals from the *Titanic* – and the question asked in the book is the obvious one; why was nothing done to assist in the disaster?

The novel that described a liner named *Titan* sinking after hitting an iceberg, published before the *Titanic* disaster.

The *Titanic* was not the only vessel to feature in an epic disaster movie. *The Poseidon Adventure* (1972), based on a novel by Paul Gallico, was another box office triumph. The ageing liner *Poseidon* is hit by a tsunami and is capsized, leaving passengers trapped in the upturned hull. The story deals with the attempts of one group to find a way to safety. A sequel, *Beyond the Poseidon Adventure*, attempted to cash in on the success of the original, but proved a dismal failure.

Turning from disaster to comedy, one of the first movies to feature a trip on a liner was *The Navigator* (1924) directed and starring the great comic actor Buster Keaton. Due to a series of errors and mishaps, he finds himself on the ship 'the *Navigator*' cast adrift in the Pacific – and, as he later discovers, with just the one woman he had hoped to marry also aboard. Much of the comedy derives from the fact that ships like this are

Members of the cast of *The Poseidon Adventure*, celebrating winning Oscars on part of the original set.

not designed to be run by a crew of two – in one scene, for example, he has to use a machine intended to boil eggs for a boat load of passengers just to make breakfast for himself. It is, in this author's opinion, not just one of the finest of silent screen comedies but one of the best, whether silent or with sound. The comic possibilities of life on board a liner were taken to extremes of mayhem in the Marx Brothers' *A Night at the Opera* (1935) where the brothers are emigrating to America. At the opposite end of the social scale, Marilyn Monroe as Lorelei Lee in *Gentlemen Prefer Blondes* (1953) is much more interested in potential millionaires in the first class part of the liner. Her plans do not go quite as she intended, having bribed the purser to get her seated at the same table as the richest passenger on board, she finds he is a ghastly little spoiled brat, not an elderly man who might be won over by her charms. The movie also features one of Monroe's best known musical numbers – *Diamonds are a Girl's Best Friend*, which neatly sums up Lorelei's philosophy of life. *The Captain's Table* (1959) is a British comedy film, featuring John Gregson transferred from captaining a cargo ship to an ocean liner, where he finds the social life far more difficult to manage than the ship. This is a broad comedy in the Carry On tradition.

One of the attractions to authors of a ship setting is that, for the length of the voyage, it is a closed community. This makes it a

Buster Keaton puzzling out how to boil an egg on a ship that is only equipped to boil them by the hundred.

good choice for the mystery writer. One author who has taken on the challenge is Conrad Allen – also rather confusingly sometimes appearing as Edward Marston. He has written a series of detective stories based on genuine ocean liners, beginning with *Murder on the Lusitania* published in 1999. Apart from having good and intriguing plots, the books give an accurate, well-researched view of life on the great liners in their heydays. Another novel set in the early years of ocean travel, *Crossing on the Paris* (2012) by Dana Gynther, is set in the 1920s and tells the interlocking story of three women, one working on the ship and travelling steerage, the next in second class and an elderly lady in first class. This is an upstairs-downstairs at sea account that contrasts the different social classes and conditions of the time.

Moving on to the 1930s and we reach a more sombre time, and few stories are more sombre than *The German Girl* (2017) by Armando Lucas Cornea. It is based on a real life story of the voyage of the SS *St. Louis* in 1939. The war in Europe had not yet begun, but the persecution of Jews in Germany had and those who could get away did so. On 13 May 1939 nearly a thousand Jews, many of them children, embarked on the ship, with visas they had bought which entitled them to settle in Cuba.

Although the Captain 'Gustav Schroder' was German, he was clearly not a Nazi sympathiser, as he insisted that the Jewish passengers should be treated with as much respect and share the same facilities as the rest of the passengers. He even took down Hitler's portrait in the dining saloon when they gathered for Friday prayers. They reached Cuba and were ready to disembark, but no one was allowed to leave. The ship remained there for seven days, while negotiations went on with the authorities. Eventually it became clear that Cuba would not be allowing them to land, even though they had paid Cuban embassy staff for their visas. The ship set sail again, this time heading for the Florida coast, but once again the authorities proved adamant. The refugees could not land in the United States. There was no longer another option. The *St. Louis* returned to Europe and although they did not have to go to Germany, many were settled in France, Belgium and Holland and when those countries were occupied the vast majority of those who had thought they were heading to safety ended their lives in the death camps.

Another novel set in the same period is *The Ocean Liner* (2018) by Marius Gabriel. This too is also based on historic facts but pushes two stories together. The historic characters, notably Arturo Toscanini and Igor Stravinsky, did both leave Europe for America on the same ship 'the SS *Manhattan*' in September 1939. The story of it being stalked by a U-Boat and threatened with sinking is also factual, but involved the *Manhattan*'s sister ship 'the SS *Washington*'. The young German boy who shared Stravinsky's cabin is simply fictitious.

A novel describing a transatlantic passage by a liner in the Second World War.

The years after the Second World War brought a new phenomenon – instead of emigrants leaving Britain, there were emigrants moving from Asia and the Caribbean to Britain. The story of three young boys travelling from Sri Lanka by liner is told in Michael Ondaatje's *The Cat's Table* (2011). The title comes from their lowly position in the ship's hierarchy – not dining at the Captain's table but down in steerage. That does not stop them from exploring the whole of the ship, often illegally, and observing how the other half lives.

One rather surprising way in which the ocean liner has been used in fiction is in the stage musical *Anything Goes*. First produced on Broadway in 1934, it contains some of Cole Porter's most popular songs, including the title song, *Anything Goes, De-Lovely, I Get a Kick Out Of You* and *You're the Top*. The story, as in so many musicals, is complicated and decidedly unrealistic, involving among other characters, a gangster pretending to be a clergyman, a stowaway mistaken for a gangster and, of course, a complicated love affair. The book is, however, witty – not surprising when the co-author was P.G. Wodehouse. It might seem all but impossible to recreate the atmosphere of a liner on stage, but a recent production proved that with clever set design it could be triumphantly achieved. It was made into a film the following year.

This is not intended as a complete list of every book and film ever written or made set on liners but is only intended to give an idea of the rich scope for fiction that the life on board presents. It is also slightly biased, as I have picked works that I have personally enjoyed.

A fictional account of young boys travelling from Sri Lanka to Britain.

A stage production of Cole Porter's Anything Goes, set on a liner in the 1930s.

Chapter Fourteen

The End of the Line

As one by one, the great lines began to abandon their regular services, they had a problem to solve. They still had magnificent ships in service, worth far more than their scrap value. They needed to find what to do with them. The answer was obvious, convert them for use for the rapidly growing cruising market.

The *QE2* had enjoyed a remarkably long career on the Atlantic route, having made over a thousand voyages and having carried more than two million passengers, but Cunard was taken over by Carnival Operation in 1998 and the following year a major refurbishment programme began to prepare her for a new role as a cruise ship. The Royal Promenade that had once catered for the wealthy with very upmarket shops was now lined with the more mundane shops selling jewellery, souvenirs and perfumes that cruise passengers expected to find. She remained a handsome vessel with some amenities that even the best cruise ships could not always boast, including a 6,000 book library. She still continued to run between Britain and America, but also cruised the world. She was finally retired in 2008 and was bought with the intention of turning her into a floating hotel. Plans, however,

The *Queen Mary 2* at Boston.

failed to be put into practice, and over the next few years, various proposals came and went, including everything from being returned to Britain to being scrapped in China. Eventually, in 2018, she did indeed become a hotel, moored at Dubai.

Cunard, however, had not yet entirely abandoned the Atlantic trade. In 2002, work began on a new ship to be called the *Queen Mary 2*. There were to be many differences from her predecessors. Previously, Cunard ships had been built in Britain, but now the contract went to the French company Chantiers de l'Atlantique who had built the older Cunard rivals, *Normandie* and *France*. Power is supplied by four 16,800 kW diesel engines, with two gas turbines for additional power when needed. Where earlier liners had cabins below decks and public rooms at upper level, here cabins are above the public rooms. The latter are all more than generous spaces offering all kinds of entertainments, restaurants and bars. Passengers can spend time at the cinema, in the theatre or try their luck in the casino; they have a range of places to drink, anywhere from the Golden Lion pub to the Veuve Cliquot champagne bar and have a wide choice of restaurants, including Italian and Indian. If all that proves a bit too exciting, they can relax in one of the four pools, luxuriate in the spa or settle down for a quiet read in the library. If the description makes it sound more like a cruise ship than an Atlantic liner, this is because for much of the year, she does indeed cruise the world. There were to be no liners commissioned by Cunard. There was, however, to be one more Queen – the *Queen Victoria*. Launched in 2007, she looks exactly what she is – a cruise ship and very much not an ocean liner.

In 1970, the SS *United States* was sent for a regular overhaul to Newport News in Virginia when the owners announced they were withdrawing her from service, and she was moved across the James River to Norfolk. There followed various attempts to find a new role for the great ship, including a plan for turning it into a time-share cruise ship. With nothing proving viable, the splendid fittings and furnishings were sold off and in 1992 the vessel was bought by a Turkish shipping company and towed to that country, where the whole ship's interior was stripped away down to the bulkheads. In 1996, she was towed back to America, and once again various schemes for bringing her back into service in some way were brought forward and ultimately failed. It seemed one of the finest liners ever built was destined for the scrap yard, but in 2009, the SS *United States* Conservancy Group was formed with the idea of saving and preserving her. After many attempts to find an appropriate scheme, the Conservancy reached an agreement with RXR Reality to preserve the ship as a floating hotel, with a variety of other facilities, such as conference centres. Several American cities have shown an interest in

The spa room aboard the *Queen Mary 2*

having the ship moored there as a tourist attraction, but at the time of writing, no firm agreement has yet been reached. One of the early options for the *United States* was a sale to the Norwegian Caribbean Cruise Line (NCL) but when that fell through, the NCL bought one of the other great liners of the age – the SS *France.*

The *France* had been withdrawn from the regular Atlantic runs and in the 1970s began operating cruises in the Mediterranean and later round the world cruises. Things seemed to be going well, but with soaring oil prices in 1973 and the withdrawal of a government subsidy, there was little option. She was put up for sale and that was when she was bought by NCL and renamed *Norway*. In order to keep her in business as a cruise ship, in the face of increasing competition, two extra decks were added. By 2004, however, her sailing days finally came to an end and she was eventually scrapped – only the bow section was saved. It was sold at auction and is now set up as a monument to the pride of French shipping at a marina at Port de Grenelle, Paris.

P & O were in a better position than many other lines, in that they had already had extensive cruising programmes before the threat from air travel hit the industry. They had introduced cruising as early as the 1930s and used liners such as *Uganda* for off-season cruising. The challenge was also not so great on their Far Eastern services, and regular runs to India did not stop until 1982. Nevertheless, the company saw the way things were heading and in 1973 put in an order for their first purpose-built cruise ship, the *Spirit of London*. In 1974 they bought up Princess Cruises. They have continued to build the fleet and added seven new vessels to the fleet in the present century with an eighth under construction at the time of writing. In 2000, Princess International broke away from P & O and is now one of the leading cruise operators in the world and has eight vast new ships built this century. A new vessel under construction, *Sun Princess*, will be the first to be fuelled by liquified natural gas.

They are not, however, the biggest operators; that honour goes to the Norwegian company Royal Caribbean, founded in 1968. They currently have a fleet of twenty-six ships, with four more due to join by 2026. They offer a range of entertainments and pastimes that would have been unimaginable on even the grandest ocean liner. The biggest ship of the group, the *Wonder of the Seas*, for example, boasts thirteen different restaurants, four swimming pools, six whirlpools, a surf simulator and monster water slides and for those with a sense of adventure, there is even a climbing wall – and those are just a few of the things on offer. It is a new concept – the ship no longer takes you to your holiday destination – it is the holiday destination.

The new generation of cruise ships are undoubtedly technologically splendid examples of modern marine engineering, but in order to cram

The cruise ship *Ruby Princess* at Istanbul.

in not only large numbers of passengers, but an unparalleled range of activities, compromises had to be made. The ships have to be able to fit in with existing harbours in places that the passengers want to visit. That limits the length of the vessels so the only way to extend is to build upwards. The result is, to this author at least, a set of vessels that hardly look like ships at all and rather more like floating apartment blocks. The interiors may sometimes have been bizarre recreations of old world hotels, but seen out on the ocean they had, at their best, true beauty of line. The great ocean liners looked simply magnificent and had about them a unique air of romance, which is why, even when, like the SS *United States*, they have had their innards torn out, people simply want to preserve them as examples of supreme marine engineering. In spite of all the bells and whistles, it is hard to imagine anyone wanting to keep a modern cruise ship for its aesthetic qualities, once its sailing days are over. The days of the ocean liner may be gone and are unlikely ever to return, but their memory survives, in pictures, photos and film and in a few happy cases as preserved survivors as epitomes of elegant design.

The pool bar swimming pool on the Royal Caribbean cruise ship *Independence of the Seas* – just one of four pools on the vessel, together with eight whirl pools and a surf simulator.

Appendix

The blue riband of the Atlantic is an award given for the fastest average speed for crossings between Europe and North America. The following list is for the ships that won the title for the westbound route. Several of them beat their own times more than once, so the table lists the dates on which they achieved this and the fastest times they achieved.

Ship	Owner	Dates	Fastest crossing (knots)
Sirius	British & American	1838	8.03
Great Western	Great Western	1838, 1839, 1843	10.03
Columbia	Cunard	1841	9.78
Cambria	Cunard	1845	10.71
America	Cunard	1848	11.71
Europa	Cunard	1848	11.79
Asia	Cunard	1850	12.25
Pacific	Collins	1850	12.46
Baltic	Collins	1851, 1854	13.04
Persia	Cunard	1856	13.11
Scotia	Cunard	1863	14.46
Adriatic	White Star	1872	14.53
Germanic	White Star	1875, 1877	15.76
City of Berlin	White Star	1875	15.21
Britannic	White Star	1876	15.43
Alaska	Guion	1882, 1883	17.05
Oregon	Guion	1884	18.56
Etruria	Cunard	1885, 1888	19.56
Umbria	Cunard	1887	19.22
City of Paris	Inman	1889, 1892	20.70
Majestic	White Star	1891	20.1
Teutonic	White Star	1891	20.35
Campania	Cunard	1893, 1894	21.44

Ship	Owner	Dates	Fastest crossing (knots)
Lucania	Cunard	1894	21.81
Kaiser Wilhelm der Grosse	Norddeutscher Lloyd (NDL)	1898	22.29
Deutschland	Hapag	1900, 1901, 1903	23.15
Lusitania	Cunard	1907, 1908, 1909	25.65
Mauretania	Cunard	1909	26.06
Bremen	NDL	1929, 1933	27.92
Europa	NDL	1930	27.91
Rex	Italian	1933	28.92
Normandie	Companie Générale Transatlantique	1935, 1937	30.58
Queen Mary	Cunard-White Star	1936, 1938	30.99
United States	United States Line	1952	34.51

Further Reading

Anderson, Roy Claude, *White Star*, T. Stephenson & Sons Ltd, 1964

Brinnin, John, *The Sway of the Grand Saloon*, Arlington Book Publishers Ltd, 1971

Burton, Anthony, *The Rise and Fall of British Ship Building*, Constable, 2013

Coleman, Terry, *Passage to America*, Pimlico, 1972

De Kerbrech, Richard P., *Ships of the White Star Line,* Ian Allan, 1955

Dorman, Frank E., *Ships of the Cunard Line,* 1955

Edington, Sarah, *The Captain's Table,* National Maritime Museum, 2005

Fowler, William M., *Steam Titans: Cunard, Collins and the Epic Battle for Commerce on the North Atlantic*, Bloomsbury USA, 2017

Fry, Henry, *The History of North Atlantic Sea Navigation,* 1896

Maber, John M., *Channel Packets and Ocean Liners, 1850-1970,* National Maritime Museum, 1980

Rentell, Philip, *P&O – Orient Liners,* Waterfront Publications, 1990

Acknowledgements

The author wishes to thank the following for permission to reproduce photographs. Other illustrations used are from the author's collection or are in the public domain.

Alex Mavel, 178; Australian National Maritime Museum, 128; Ballista, 35; Benjamin B. Hampton, 171; Brunel University, 59; Business Records Centre, University of Glasgow, 63; C. Hullmandel, 26; Colin Rose, 68; David Jones, Francis Browne, 139; Glasgow Museums, 51; Gwynedd Archives, 50; Hannes Grobe, 140; John Ward, 138; Jolly Janner, 178; Library of Congress, 21,147; N.E.I. Parsons Ltd, 115; Memorial Student Centre, Texas, 174; Nikolay Checkanov, 162; NYPL, 43; Ocean History 46, 174; Perchristener, 75; Press Agency, 161; Samuel Walters, 28; Scottish Records Office, 53; Trondheim Havn, 161; Tyne & Wear Archives, 62.

Index